RECENT ADVANCES IN EMBRYOLOGY

RECENT ADVANCES IN EMBRYOLOGY

Vol. 1
Embryology of Fishes

A.P. Diwan
&
N.K. Dhakad

ANMOL PUBLICATIONS PVT LTD
New Delhi-110 002

ANMOL PUBLICATIONS PVT LTD
4374 / 4B, Ansari Road
Daryaganj, New Delhi-110 002

Recent Advances in Embryology

First Edition 1994
ISBN 81-7041-974-3 (Set)

PRINTED IN INDIA

Published by J.L. Kumar for Anmol Publications Pvt. Ltd., New Delhi
and Printed at Efficient Offset Printers, Delhi.

Preface

Embryology is a branch of biology which has a most immediate bearing on the problem of life. It has become an immense field whose confines are difficult to delineate. The focus has been and remain on the embryos, on the gradual emergence of form and structure from what appears to be a very modest beginning. The essence of embryonic development is change transition from one stage to another. Embryo is a fleeting stage, a continum along the axis of time. Embryology is a bank division of biology. It stems the anatomy, the histology and the physiology of adult. To understand it well is to aid in the comprehension of the other biological disciplines.

Though, the authors have taken special care to present a current account, yet they are fully aware about their limitations, and the readers may come across the mistakes of various types. For all types of mistakes they extend their due apology.

The authors have freely consulted other standard books and research papers while preparing the manuscript, so the authors claim no originality of the work.

The authors express their thanks to their friends and colleagues whose continuous inspirations have invited them to bring out this text.

The authors are also thankful to Shri J.L Kumar of M/s Anmol Publications Pvt. Ltd. for bringing out a handsome print of this set in a comparatively short time.

Constructive criticisms and suggestions for improvement of the book will thankfully acknowledged.

Authors

Contents

1

Reproduction in Fishes

Vertebrates do not reproduce asexually. The fishes, like all the other vertebrates are known to reproduce sexually. Being unisexual, eggs and spermatozoa are formed in separate individuals *(dioecious)*, and the gametes are expelled into the surrounding water where fertilization takes place immediately and is promptly followed by the development of a new generation. Although most fishes are dioecious, hermaphroditism does occur-particularly among the cyclostomes and teleosts; in some species it is the normal way of life. *Parthenogenesis* which is the development without fertilization has not been observed among fishes in nature but has been produced experimentally. *Gynogenesis* occurs in *Poecilia formosa* and perhaps also in some populations of *Carassius curatus*. These modes of reproduction are, however, unusual, and most of the 20,000 or more known species of fish have male and female organs in separate individuals (dioceous).

At one extreme, the two sexes are externally indistinguishable. They swim together during the breeding season to discharge their genital products into the water without specialized mating behaviour. At the other extreme, marked differences in morphology and colouration are characteristic of the males and females and these play distinctive roles in elaborate presexual behaviour, courtship, mating and parental care. A synchronized spawning without copulation is usual but copulation occurs at all levels in fish phylogeny. It sometimes merely insures close proximity of eggs and sperm discharged into the water but more frequently involves insemination of the females. Although internal fertilization may be followed by the laying of newly fertilized eggs or the release of primitive larvae, viviparity is highly specialized in some elasmobranchs and teleosts with young born in advanced stages of development; in *Cymatogaster aggregata* the males may even be sexually mature at birth. Sexual activities are normally followed by the prompt fertilization of eggs but sometimes sperm are stored in the female for prolonged periods prior to fertilization; occasionally the fertilized eggs develop only after a period of diapause; sometimes development occurs in a usual way but hatching is delayed due to unfavourable conditions.

Reproductive Organs

The gonads of all vertebrates develop in the dorsolateral lining of the peritoneal cavity — one gonad on each side of the dorsal mesentery. Their development is intimately associated with that of the excretory system. In most of the vertebrates each gonad has a double origin, developing from two distinct but closely associated cellular proliferations. The more laterally located cortex or cortical portion arises is destined to become an *ovary*. The *meddula* or medullary portion which is destined to form the *testis* arises from a more medial cellular proliferation which also produces the adrenocortical tissue. Usually, one of these portions grows rapidly while the other fails to develop and the sex of the individual is thus determined at a very early stage. This pattern of gonad differentiation from two different components is characteristic of the elasmobranchs and all of the tetrapods. In contrast, the gonads of cyclostomes and teleosts develop from single primordia. In cyclostomes and teleosts the entire gonad develops directly in the peritoneal epithelium and corresponds to only the cortex of other vertebrates. There is evidently no contribution from the interrenal blastema.

Whether the ontogeny of the gonad is by way of a single or a double primordium, distinctive cells (*the primordial germ cells*) which are destined to form gametes first appear within or migrate into the cortical portion of the gonad. They can

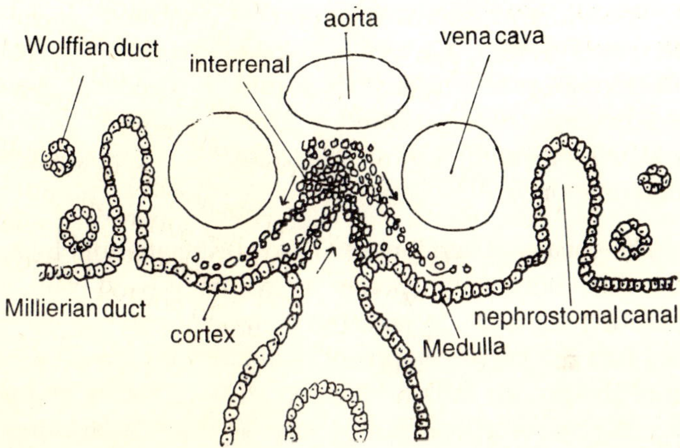

Fig. 1.1.   Showing the origin of cotex and medulla of the gonad in dogfish.

usually be identified as conspicuously larger cells within this modified layer of proliferating germinal epithelium. There is a voluminous literature on the origin of the germ cells with a considerable body of evidence for a widespread development of these cells and a subsequent migration of them into the germinal epithelium. In genetic male elasmobranchs there is an early migration of germ cells from the cortex, where they first appear, into the medulla. In the genetic females there may be a transient migration of some of the germ cells into the medulla but many of them retain their cortical location to form the basis of ovarian differentiation.

Under normal conditions, genetic factors probably determine whether the developing fish will be a male or female. Earlier workers postulated male and female inductor substances or hormones which controlled the course of development. *Witschi* (1942, 1950) who carried out the pioneer work on amphibians referred to them as *"medullarian"* and *"corticin"*, for male and female, respectively; *D'Ancona* (1945, 1950) working with the teleosts called these theoretical substances *"androgenin"* and *"gynogenin."* *Reinboth* (1962) does not consider evidence for the existence of these factor to be at all convincing. There are special embryonic hormones concerned with the determination of sex. It is well established that the differentiation of the gonadal primordium can be readily modified with gonadal steroids similar to those found in adults. *Androgens* stimulate the development of testes, and *estrogens* promote ovarian differentiation.

The gonads of vertebrates always originate from bilateral primorida, but many species as adults possess only one reproductive gland. In lamprey there is a fusion of the two primordia during development, while in the case of Myxine one of the gonads fails to develop. The literature reveals a range of specializations in all groups of fishes from complete fusions to partial fusions involving only the posterior portion of the gonads or just the gonoducts; sometimes one of the gonads is rudimentary or merely smaller but still present.

Some most interesting problems in the phylogeny of the gonoducts and their relationships to the mesonephric tubules and ducts have been observed. Gonoducts are absent in the cyclostomes. Spermatozoa or ova are discharged from the surface of the gonad directly into the body cavity and then pass through pores into the urinary sinus or urinary duct from where they are expelled through the cloaca or urinogenital papilla — depending on the anatomy of the urinogenital opening in the particular sex and species.

Gonoducts are present in all groups of the gnathostomes although they may be secondarily lost in some fishes like the Greenland shark, *Laemargus borealis*, and the Salmonidae among the Teleostei, Kerr provided the classic description of the origin of the male gonoducts from the mesonephric system. The sturgeon and garpike are thought to represent a primitive situation where some of the renal tubules throughout the length of the mesonephros have been conscripted into the service of the testis and form vasa efferentia which drain into the mesonephric duct or vas deferens. In the Chondrichthyes and Amphibia, the testis is thought to have taken over a group

of anterior mesonephric tubules which cease to have any
relationship with the excretory system; in *Lepidosiren*, the vasa
efferentia are formed from some of the posterior mesonephric
tubules. In *Polypterus* and the Teleostei there is no connection
between the mesonephros and the gonad at maturity; and the
vas deferens is quite separate from the ureter or mesonephric
duct. It is generally assumed, however, that the main gonoduct
has been derived from the mesonephric duct during phylogeny.

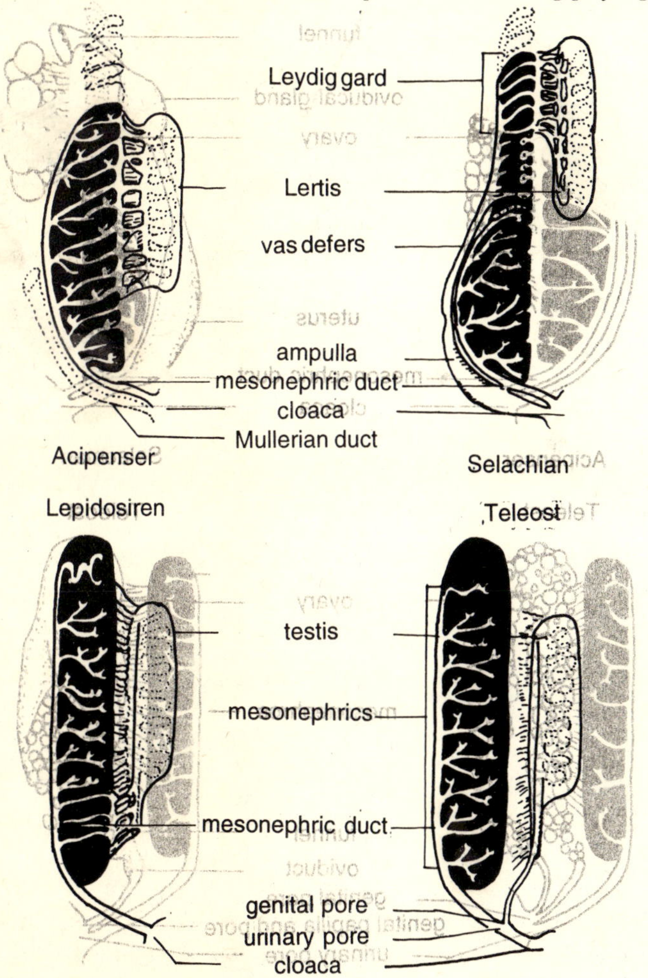

Fig. 1.2. Urinogenital organs in male fishes.

In almost all vertebrates, except some of the more specialized fishes, the ova are discharged into the peritoneal cavity and find their way to the outside through oviducts, which pass from open anterior funnels to the cloaca. In these groups with naked ovaries (*gymnovarian* condition) and open ovarian funnels, the genital ducts are derived as in the male from the mesonephric ducts although the evidence for this origin is completely lost in the land vertebrates.

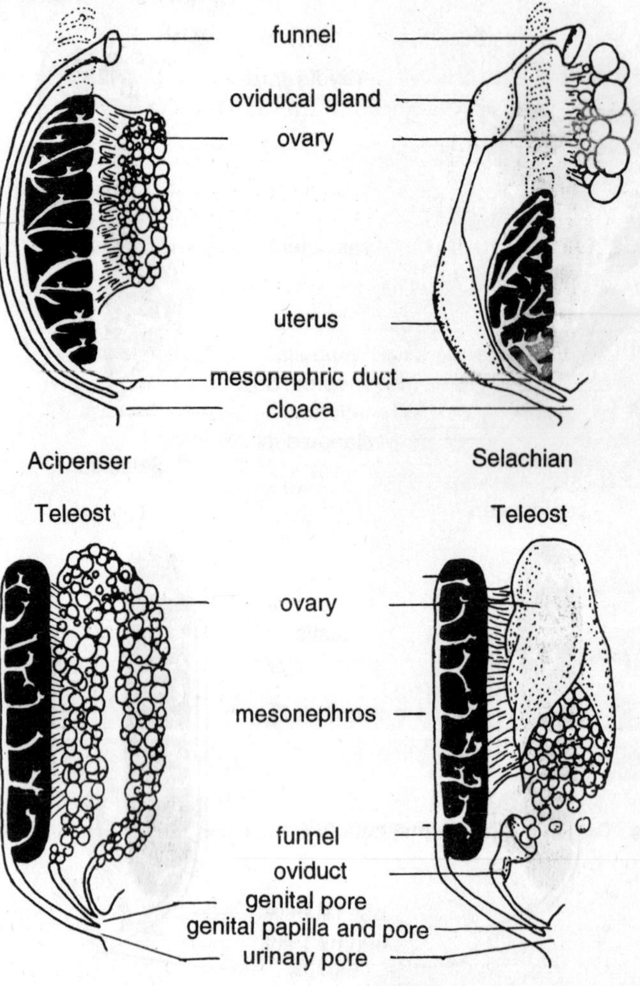

Fig. 1.3.   Urinogenital organ in female fishes.

In fishes belonging to the group Teleostei, the oviducts are posterior continuations of the ovarian tunic. The embryology of the ovary and its duct varies so that the ovary in some teleosts has a central ovarian cavity continuous with the oviduct while in others the oviducts are para-ovarian; in any case the oviducts are formed by the backward growth of the same peritoneal folds which enclose the ovary during its development (*cystovarian* condition). The gonoducts are also continuous with the ovaries in the holostean *Lepidosteus*, thus providing the exception to the rule that only teleosts fail to discharge their ova into the peritoneal cavity. Some of the teleosts are themselves exceptional in that they do release their eggs directly into the body cavity. In some cases the oviducts degenerate in whole or in part so that the ova pass into the peritoneal cavity and thence through pores or funnels, depending on the degree of degeneration, to the exterior. In the Salmonidae, Galaxiidae, Hyodontidae, Notopteridae, Osteoglassidae, and the cyprinid *Misgurnus*. In the Anguillidae the loss of gonoducts occurs in both males and females.

THE MALE REPRODUCTIVE ORGANS

The Testis and Spermatogenesis

The process forming spermatozoa from the sperm mother cells or spermatogonia through a series of cytological stages collectively referred to as "*spermatogenesis.*" This process involves a proliferation of spermatogonia through repeated mitotic divisions and growth to form diploid *primary spermatocytes;* these then undergo reduction division to form haploid *secondary spermatocytes;* the division of the secondary spermatocytes produces the haploid *spermatids* which then metamorphose into the motile and potentially functional gametes — *spermatozoa, spermia* or *sperm.* This process of spermatid metamorphosis is often called "*spermiogenesis.*" Both environmental and hormonal factors trigger waves of spermatogenesis at different seasons and control the essential steps of meiosis i.e., division of primary to secondary spermatocytes and the metamorphosis of the spermatid with eventual release of mature sperms. In some species—particularly the elasmobranchs and viviparous teleosts—sperm production involves the packaging of sperms into sperm balls or *spermatophores* which are transferred to the female during courtship or mating.

The process of *spermatogenesis* occurs within testicular units which may take the form of small sacs, ampullae, lobules, or tubules; in many groups of fishes these differ radically from the familiar seminiferous tubules of the mammalian testis. In the cyclostome, spermatogenesis occurs within *small bladders, follicles, or ampullae.* These are separated by a delicate connective tissue. A number of units may be grouped together and bounded by somewhat thicker connective tissue to form *lobules.* Spermatogenesis is almost synchronous throughout the many follicles and just prior to spawning the follicles filled with mature sperm rupture to release their contents into the body cavity.

In fishes like elasmobranchs, *spermatogenesis* occurs within a mass of *ampullae* arranged in a manner which seems to be unique among the vertebrates. The testis of the basking shark, *Cetorhinus maximus,* is divided by connective tissue trabeculae into many lobules each of which corresponds to the entire testis of the dogfish, *Scylliorhinus canicula.* The spermatogenetic units, usually called *"ampullae,"* are proliferated from a mesoventral area of the testis referred to as the *"tubulogenic zone."* Within this zone, nests of cells — somewhat like primary ovarian follicles — arise and proliferate to form small tubules or ampullae which gradually shift toward the dorsal side of the organ while spermatogenesis occurs within them. By the time the ampullae reach the dorsal surface of the testis, the sperms are ready for discharge into the efferent ducts which emerge from the testis at this point. At this stage, the *Sertoli cells* surround the ampullae and are clearly associated with packets of sperms. According to L. H. Matthews (1950), the ampullae shrink after sperm are discharged into these collecting tubules and the *Sertoli cells* are resorbed. It is of interest that

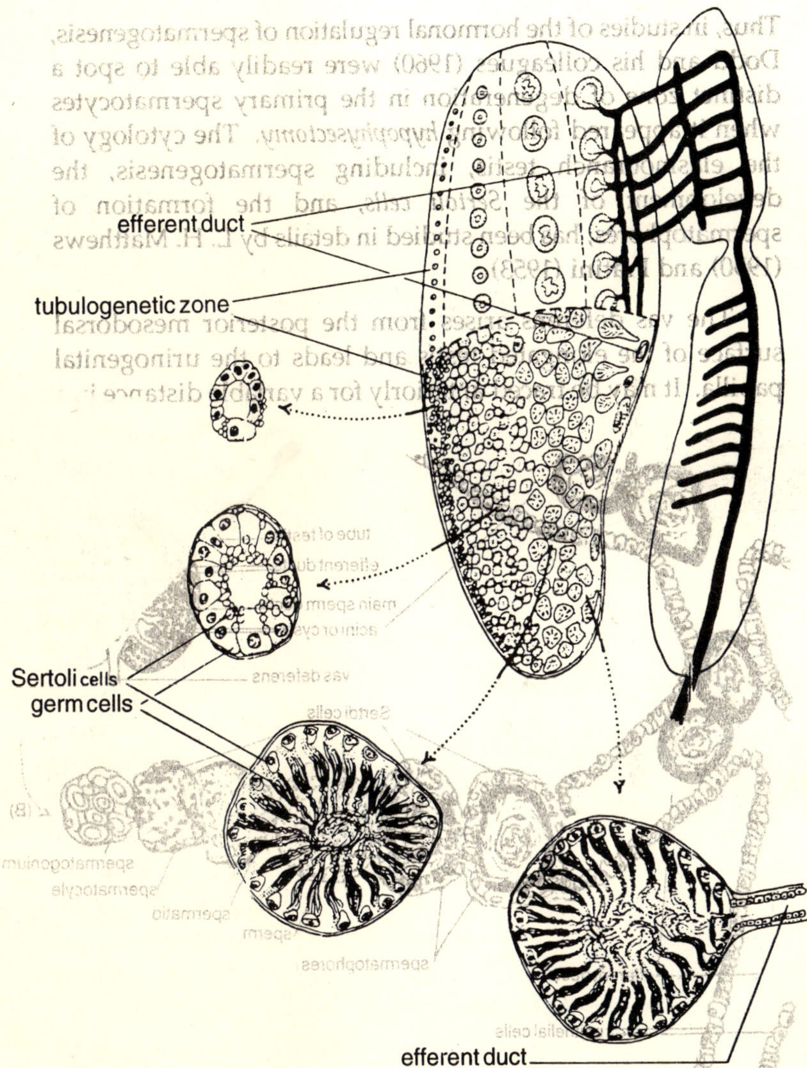

Fig. 1.4. Structure of the testis of a Chondrichthyan fish.

all of the gonocytes within any one ampulla are in the same stage of spermatogenesis and that within the testis, distinct zones are evident from ventral to dorsal surface with all the tubules of a particular zone in a similar stage of development.

Thus, in studies of the hormonal regulation of spermatogenesis, Dodd and his colleagues (1960) were readily able to spot a distinct zone of degeneration in the primary spermatocytes when it appeared following *hypophysectomy*. The cytology of the elasmobranch testis, including spermatogenesis, the development of the *Sertoli cells*, and the formation of spermatophores, has been studied in details by L. H. Matthews (1950) and Fratini (1953).

The vas deferens arises from the posterior mesodorsal surface of the elongated testis and leads to the urinogenital papilla. It may be traced anteriorly for a variable distance in a

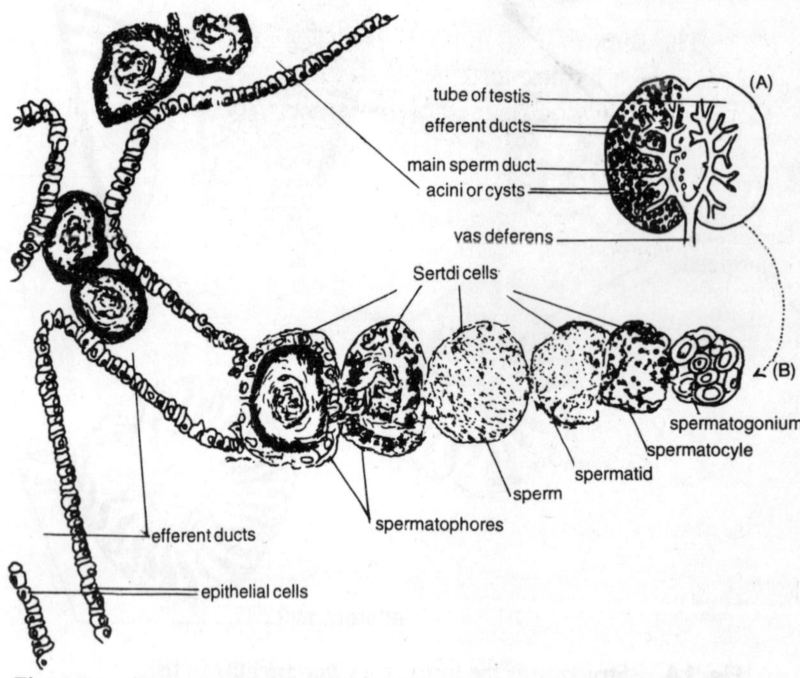

Fig. 1.5.   Structure of testis of guppy, *Poecilia reticulata*. (A) Diagrammatic section to show relation of gametogenetic tissue (acini or cysts) to the system of ducts, and (B) series of cysts to show differentiation of gametes from spermatogonia (at right) to mature spermatophores surrounded by Sertoli cells (at left). Sertoli cells become confluent with the epithelial cells of the efferent ducts when the spermatophores enter the ducts.

connective tissue groove of the testis along with the blood vessels and nerves supplying testis. In many teleosts, the paired testes fuse posteriorly and the vasa deferentia are combined into a single sperm duct. Within the body of the testis, the main sperm ducts give rise to smaller ducts the *vasa efferentia* which penetrate ventrally and laterally to form a drainage system of variable complexity. In some species these tubules are extremely short while in others they form an extensive system of seminiferous tubules which can be followed almost to the periphery of the organ. Testes of the poeciliid type are sometimes referred to as *"acinar"* while those with the extensive duct systems are called *"tubular."*

The seminiferous tubules of the teleost — in contrast to those of the higher vertebrates — lack a permanent germinal epithelium. Whether the testis is *acinar* or *tubular*, nests of spermatogonia proliferate from the resting germ cells near the margin of the organ. In the *acinar type*, these nests of cells or cysts undergo the various stages of maturation and then after are displaced toward the sperm ducts into which they eventually discharge their contents. In the *tubular testis*, the resting germ cells are particularly evident and packed together at the blind ends of the tubules near the periphery, but many of them migrate or are displaced along the walls of the tubules. In active spermatogenesis, nests of spermatogonia proliferate both from the ends of the tubules and from the resting germ cells along their cells. Thus, at the end of spermiogenesis, the seminiferous tubules are packed with sperms as the masses of gametes from a multitude of matured cysts combine within the tubules. During maturation all of the cells within one of the cysts are in approximately the same stage of development; the degree of synchrony among the many cysts varies in different species.

The Endocrine Tissues of the Testis

Well-vascularized clusters of cells similar to those described by *Leydig* many years ago in the mammalian testis have been identified between the seminiferous alevoli and tubules of many different fishes. In routine H and E sections, these large cells with spherical or oval nuclei often appear vacuolated because of the removal of lipoidal substances. Their endocrine nature was first postulated from the marked seasonal proliferation which occurs just prior to the breeding season in the stickleback *Gastterosteus* in this same species *Gottfried* and *van Mullem* (1967) have recently developed a convincing correlation between the histological development of the tissue and its biochemical content of *androgen*. Nests of typical *Lydig cells* or *interstitial tissue* have now been identified in the cyclostomes in all groups of elasmobranchs, in the lungfishes and the coelacanth *Latimeria* as well as the testes of many teleosts like *Gasterosteus Tilapia, Tinca, Solea, Lebistes, Cymatogaster, Oncorhynchus* and others. Staining of the cytoplasmic droplets with sudanophilic colouring agents and the demonstration of steroid dehydrogenases leaves little doubt of their role in *steroidogenesis*.

In some species of teleosts, however, the hormone-producing cells are located in the basement membrane of the seminiferous tubules. *Marshall* and *Lofts* (1956) who first noted this difference in their studies on the pike, *Esox lucius,* and the char, *Salvelinus willughbii,* referred to this tissue as the *"lobule boundary cells."* The difference between interstitial and lobule boundary cells is largely one of distribution since both tissues arise from the same source and are similar histochemically.

The nutritive or *Sertoli cells* are also prominent in the testes of all groups of fishes. The spermatogenetic units — whether cysts, ampullae, or tubules — are bounded by a thin layer of connective tissue : the basement membrane and contain two types of cells; one of this is the *gonocyte* giving rise to the several generations of spermatogenetic cells, which the other is the *Sertoli* or *supporting cell* believed to play a nutritive role during *spermiogenesis*. The process whereby spermatids become embedded in the centripetal end of the *Sertoli cells* to undergo metamorphosis is well described in the higher vertebrates

(Nelsen, 1953; Patten, 1953; Ham, 1965). Cytological details probably vary in the different species, but in all cases the spermatids become intimately associated with these nurse cells are presumably draw nourishment from them during transformation. The fully developed sperms are attached to the *Sertoli cells* prior to *spermiation*; this association can be very nicely seen in fishes such as the sharks, the poeciliids, and the embiotocids which form *spermatophores*. In these groups, where all of the sperms in a unit mature at the same time, the *Sertoli cells* form an almost complete layer just inside the basement membrane in many other fishes the situation is similar to that of the higher vertebrates where *Sertoli cells* are interspersed at intervals along the basement membrane with the groups of spermatogonia between them.

The Sertoli cells are nutritive as well as phagocyte as they may be concerned with the phagocytosis of unused sperms and involved in hormone production. The presence of *hydroxysteroid dehydrogenases* has been demonstrated in the *Sertoli cells* of dogfishes and in the surfperch. In summary, the cytology of the Sertoli cell suggests three functions; nutritive, phagocytic, and hormonal production and resemble the granulosa cells of the ovarian follicle.

Secretions of the Sperm Ducts

As soon as the spermatogenesis is over the mature sperms are made ready for discharge or *spermiation*. The glandur epithelial lining of the sperm duct probably always contributes to the seminal discharge but there seems to be no systematic study of this. However, a wide variation is recognized — at one extreme, the formation of specialized packets of sperms or *spermatophores* and, at the other, a mere thinning of the *semen* through the hydration of the testis and accumulation of fluid in the testicular passages.

In the basking shark, *Cetorhinus maximus*, the sperm packets range in size from a few millimeters up to 25 or 30 mm and have a cortex of translucent hyaline material surrounding a central maass of opaque white sperms; the sperm mass may be 10 mm in diameter. The general process seems to be representative of the elasmobranchs although the structure of the spermatophores varies in different species from simple sperm aggregations to the hyaline packets of the basking shark. In the elasmobranchs, sperms released from the ampullae into the efferent canals pass through a mass of coiled glandular tubules or *gland of Leydig* which are derived from the anterior nonurinary portion of the mesonephros. Sperms contained in the secretions of the *Leydig gland* then pass into an expansion of the vas deferens known as the *ampulla*. As they move through the complex system of septae in this structure they are consolidated and receive additional secretions such as the *hyaline cortex* of the basking shark spermatophore. Spermatophores are also regularly formed in the viviparous teleosts. In these fishes the aggregations of sperms formed within the seminiferous acini or tubules become arranged with their tails directed centrally and the heads oriented peripherally to form the *sperm balls*. As these pass through the efferent ducts, they seem to receive a gelatinous secretion which binds them together so that they remain intact during transfer to the female. More complex spermatophores have also been described in teleosts, but there are very few studies of either the cytogenesis or the physiology of fish spermatophores.

Spermiation in the goldfish is typical of the much simpler process involving only a thinning of the semen. *Clemens* and his associates (1964; *Clemens* and *Grant*, 1964, 1965) have described the weight changes in the testis and established a pituitary regulation of the spermiation process. *Yamazaki* and *Donaldson* (1968a) have used the *spermiation* of the goldfish in the bioassay of salmon *pituitary gonadotropin*.

In some teleosts (Ariidae, Gobeiidae, and Blennoidae), large structures often referred to as *"seminal vesicles"* are found as glandular developments from the sperm ducts — occasionally from the testis as in *Tachicorystes*. These *"seminal vesicles"* do not store sperms and are not comparable to the structures of the same name in the higher vertebrates. They provide secretions which are of importance in sperm transfer or other breeding activities. Secretory activities of the sperm ducts, accessory glands, and secondary sex characters show a marked seasonal development and are under the control of the *androgenic* secretions of the testis.

THE FEMALE REPRODUCTIVE ORGANS

Morphologically the fish ovary ranges from an expanded mesentery the *mesovarium* which dehisces mature ova from its ventral margin in the *hagfish Myxine* to a complex hollow organ in the viviparous teleosts where the gonad produces eggs, stores sperms, serves as a site for fertilizations, and provides nourishment for the development of young to an advanced stage.

The germinal epithelium which covers the surface of the ovary as an extension of the peritoneum the *mesovarium* develops the ovarian follicles. This germinal epithelium also lines the cavity of the hollow teleost (*cystovarian*) ovary. In the basking shark, *Cetorhinus maximus* the ovary is a single organ and is evidently exceptional among the elasmobranchs in that the germinal epithelium invaginates to form a series of tubular ramifications within a gonad which is superficially similar to that of the teleosts. It differs, however, both in its embryology and in its gonoduct. The cavities of the basking shark ovary open into a pocket on the right side of the organ and ova discharged into this pocket pass via the peritoneum into the open end of the *Mullerian duct*. These hollow ovaries of the teleosts and the basking shark are unique among vertebrates and quite different from the hollow ovaries of some other elasmobranchs and the amphibians, where the lining is not germinal epithelium and where the cavities develop as large lymph spaces within the stroma or *medulla* of the gland.

All the ovarian follicles of the fish ovary are supported by a richly vascular connective tissue *stroma* which extends into the gland from the somewhat denser connective tissue layer the *tunica albuginea* just under the germinal epithelium. The internal lining of the hollow teleost ovary is thrown into a complex series of folds the *ovigerous folds* which may almost obliterate the cavity. Fish eggs are discharged from mature ovarian follicles into the peritoneum or into the cavity of the ovary by a process of *ovulation*. The stroma of the ovary is rich in elastic tissue and smooth muscle to make the process easy & possible.

As the oogenesis progress further, the oogonia which arise from primordial sex cells either in or near the germinal epithelium become surrounded by a layer of small epithelial cells to form the *ovarian follicle*. In cyclostomes and teleosts this follicular epithelium is single-layered while in elasmobranchs and amniotes it is usually composed of several layers. The connective tissue near this nest of cells forms a distinct *theca*, which in some species assumes a very active role during the later history of the follicle. As the follicle differentiates and the ovum becomes mature, the epithelial cells increase in size and number to form a *glandular granulosa* while the *theca* becomes more distinct and may be differentiated into an *interna* and an *externa*. The maturing ovum is separated from the granulosa by a noncellular membrane usually called the "*zona pellucida*.

In fishes the functions of the follicular epithelium are still problematic. Like *Sertoli cells* the follicular cells are nutritive as well as phagocyte. The granulosa has a recognized responsibility for the deposition of yolk in the developing ovum and for its removal in ova which degenerate before ovulation. Yolk deposition in some elasmobranchs and reptiles apparently takes place through specialized protoplasmic processes which can be seen to penetrate the zona pellucida from particular granulosa cells; nutritive transfer is not microscopically evident in other fishes. The phagocytic activities

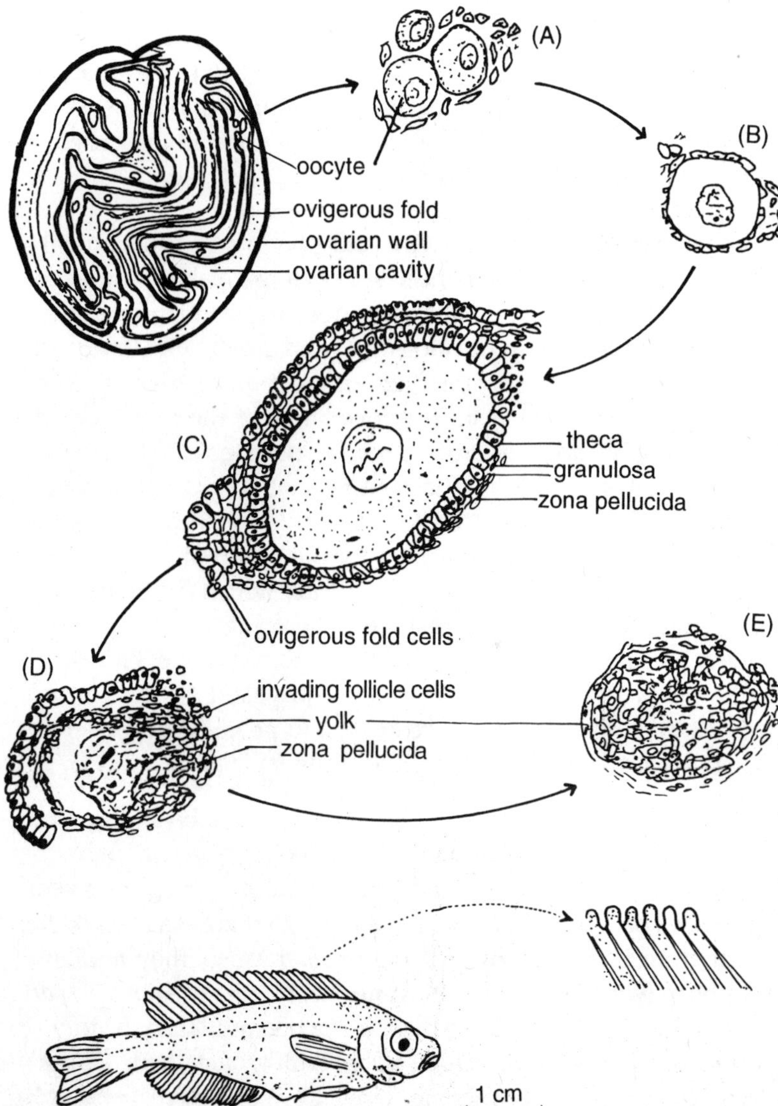

Fig. 1.6. Structure of the ovary in *Cymatogaster*. Upper left, section of ovary to show oocytes developing in ovigerous folds : (A) nest of oocytes in very early stages of development, (B) Class I oocyte, (C) Class III oocyte, (D) Stage II atretic oocyte, and (E) atretic oocyte almost at Stage III. Bottom, *Cymatogaster* before birth to show specialized dorsal fin.

of the granulosa cells have been described repeatedly by histologists since the last century. In all groups of vertebrates, the number of ovarian follicles started on the road to development is greatly in excess of the number of eggs which are eventually produced. Some ova are normally resorbed in different stages of development.

In addition to its nutritive and phagocytic functions, the granulosa may also be concerned with the elaboration of the *ovarian hormones*. The fish ovary does not contain interstitial tissue comparable in development and histochemistry to the *Leydig cells* of the testis; the theca of the follicle, which in some mammals participates in the formation of the *corpus luteum* and probably secretes *progesterone* — perhaps also *estrogen* — never shows a sudanophilia in fishes and is evidently a simple fibroelastic connective tissue. It is evident that the ovarian hormones must be synthesized by the ovum or the *granulosa* or the *corpus luteum* or *corpus atreticum* which develops from the granulosa. Major interest the centered around the corpora, which in this review are termed *"corpora lutea"* even though they may not be physiologically equivalent to the corpora lutea of mammals which are known to produce progesterone and are under the control of the pituitary gland.

In all groups of fishes *Corpora lutea* have been observed. Their histogenesis has been described repeatedly since the beginning of the century and a rich literature is cited in several reviews of fish endocrinology. Although there is a remarkable variation in the extent of these proliferations, they regularly appear when yolky ova become atretic *(preovulatory corpora lutea or corpora atretica)* or during the postovulatory history of the follicle *(postovulatory or corpora lutea of ovulation)*. The two opposing views concerning their physioly have been ably maintained from the studies of elasmobranchs by Hisaw (1959, 1963) who argues that they are concerned with yolk phagocytosis in the preovulatory follicles or the removal of tissue fragments and blood cells in the post-ovulatory ones, while Chieffi (1961,

1962, 1967) finds substantial histochemical evidence for *steroidogenesis* and attributes an endocrine function of them. Their function is just as problematic in other groups of fishes. The controversy was detailed recently and will not be repeated here; the conclusion reached at that time. (Hoar, 1965a) still seems valid; it is suggested "that estrogen synthesis was one of the responsibilities of the granulosa from the earliest stages of vertebrate phylogeny and that this capacity developed in association with the synthesis of lipid materials present in the yolk; the high estrogen content of yolk in the ova of some species may represent their entire reserve of this hormone. With the variety and complexity of reproductive processes and controls found among fishes, it seems entirely reasonable that the same granulosa cells - (in some fishes) may have specialized in hormone production to the point where *corpora atretica* become functional *corpora lutea* — even though they may synthesize *estrogen* rather than *progesterone* as their hormone."

Accessory Reproductive Structures

In female fishes, secondary sex characters are inconspicuous. The females are frequently larger, occasionally have specialized ventral fins, distinctive ventral surfaces or folds of unknown function on the abdominal wall, tubercles on the head, or an enlarged urinogenital papilla. The genitalipapilla may be only slightly enlarged but in the cyprinid *Rhodeus amarus* — the European bitterling — it often extends well beyond the caudal fin as a specialized ovipositor.

Physiologically the oviduct have been almost completely confined to the Chondrichthyes. In this group, the oviduct which is the Mullerian duct not only serves as an open tube for the collection of eggs from the abdominal cavity and their transport to the cloaca but also provides the secretions concerned with the formation of horny shelled eggs in the oviparous species and a site for the development of the young in the viviparous forms. In addition, the oviduct also serves for the reception of sperms and, at least in some species, for sperm storage and the dissolution of the hyaline cortex of the spermatophores.

The oviduct of viviparous teleosts is also concerned with some of these sperm functions, but there are no systematic physiological studies.

The Mullerian duct is differentated into four regions the *ostium* or *funnel*, the *oviducal* or *nidamental gland*, the *connecting isthmus*, and the expanded posterior *uterus*. In some species the ostium is closely applied to an ovarian pocket from which the ova emerge to enter it directly; in other species the eggs are discharged at many points on the surface of the ovary and carried into the funnel by continuously beating cilia which line the abdominal cavity. The ciliation of the abdominal cavity has been described and illustrated by Metten (1941). The absence of cilia in males and immature females and changes in the size of the ostium during the breeding season suggest a hormontal regulation of these structures; *estrogens* have been shown to stimulate the development of *nidamental glands* and other areas of the Mullerian tubes.

The oviducal gland in all chondrichthyans secretes albumen and mucus; it is relatively larger in the oviparous species where it is also responsible for the formation of the shell. Two or three distinct glandular zones may be distinguished. In the oviparous species there is an anterior *albumen-secreting area* and a posterior *shell-secreting zone;* an intermediate *mucus secreting area* may be present, as in the ratfish, *Hydrolagus colliei,* or absent, as in the ovoviviparous species *Rhinobatus granulatus* or located caudad as in *Raja rhina.* The shell-secreting area of the nidamental gland also serves as a seminal vesicle in the dogfish where sperms are stored to fertilize eggs before or at the time of shell formation.

The wall of the *uterus* is smooth and covered with a flattened epithelium in the oviparous elasmobranchs where it serves only as a passage-way for the eggs. In viviparous species, this portion of the Mullerian duct is variously modified through the formation of villuslike appendages which nourish the developing young.

VIVIPARITY AND GESTATION IN FISHES

Each adult member of the population must on the average produce one reproductively active adult if an animal species is to survive. At one extreme, eggs and sperms are broadcast in sufficient numbers to balance the unusual pressures of the environmental and satisfy the predators while, at another extreme, fertilization is internal and the developing young ones are housed within the parent's body until ready for an independent existence. The provision of millions of unprotected gametes represents the primitive condition among fishes. Along the road to specialized viviparity many curious and successful devices have evolved to provide a measure of protection during incubation. Parental care must confer great biological advantages in reducing energy demands for production of eggs or by spreading this load ever a longer period while the young are developing within the parent.

Between those fishes which represent the primitive situation—scattering millions of unprotected gametes — and the highly specialized viviparous forms, there are numerous species which build nests and exercise parental care (Salmonidae and Gasterosteidae), other species in which incubation takes place either in the buccal (Cichlidae and Bagridae) or branchial chambers (Amblyopsidae), fishes which provide special devices for the attachment of their young to their bodies (cutaneous incubation in the male *Kurtus gulliveri* or the females of *Aspredo cotylephorus* and *Solenostomus laciniatus*), and the fascinating sea horses and pipefishes (Syngnathidae) in which the males have a ventral marsupial pouch where the young are incubated until ready for an independent existence.

True viviparity has been observed an elasmobranchs and teleosts. In each of these groups there is an array of species from the ovoviviparous — where the eggs have sufficient yolk for the nourishment of the young and the female provides only protection — to the truly viviparous species — where the yolk content of the egg is greatly reduced and the developing young establish a connection with the maternal tissues at an early stage to draw nourishment from them and to satisfy the respiratory and excretory demands.

Viviparity Among the Chondrichthyes

The Chondrichthyes produce relatively few yolky *telolecithal* eggs which vary greatly in their content of organic matter. In the truly oviparous groups (Scylliade, Heterodontidae, Raiidae, and Chimaeridae), ova are housed in horny protective shells which provide protection throughout a long period of incubation. In the ovoviviparous and viviparious species, the young develop within the *Mullerian duct* (uterus) on which they usually depend for nourishment as well as protection.

Internal Fertilization

Fertilization is always internal among the Chondrichthyes. The pelvic fins of males are modified into the copulatory organs or *claspers*. These posterior extensions of the fins are stiffened not only by the cartilages of the metapterygia but also during copulation by erectile tissue. The relative contributions of cartilage and erectile tissues vary in different species. In all cases, an essential portion of the organ is the *clasper groove* formed by skin folds, the edges of which overlap to form a scroll-like tube along which the sperms are transported from the cloaca. Another constant feature of this apparatus is the presence of a clasper *syphon* or *gland* which contributes to the seminal discharge or provides a pumping mechanism for its release. In the sharks, the synphon takes the form of a blind muscular sac situated just under the skin anterior and lateral to the cloacal region; in the skates and rays a clasper gland takes the place of the hollow syphon. The dilute fluid found in the syphon was mostly seawater and that during copulation the contractions of the muscular syphon wall pumped sperms through the clasper groove in a jet of seawater. It is by no means certain, however, that this is its main function.

It has been reported that the syphon of the spiny dogfish, *Squalus acanthias*, secretes an abundance of 5-hydroxytryptamine (*serotonin*); a pair of syphons in a mature male may contain as much as 20 mg of serotonin or 0.3%. The further demonstration of a marked stimulatory effect of serotonin on the isolated uterus of the spiny dogfish is strongly suggestive of a role for the syphon in the transport of sperms after transfer to the female. At this stage, generalizations are not justified since only traces of serotonin have been found in the syphon sacs of the smooth dogidsh, *Mustelus canis*, and none was demonstrated in the clasper glands of *Torpedo* and *Raja*. The clasper glands of skates and rays produce a milky white semiviscous fluid which coagulates on contact with seawater.

Ovoviviparity
It is laying of some what underdeveloped yong ares. The ovoviviparous species are far more numerous than the truly placental or viviparous ones; a true viviparity with yolk sac placenta is confined to certain species of two families of sharks — the Carcharhinidae and the Sphyrnidae, both belonging to the Galeiformes. Actually, the distinction between ovoviviparity and viviparity is a rather artificial one for the physiologist since the maternal contribution of the placental shark to the nourishment of the fetus is an intermediate one in a series ranging from almost zero in the primitive ovoviviparous species to an almost complete dependence in the highly complex ones. Moreover, in some sharks (*Sphyrna tiburo* and *Mustelus canis*) the placenta develops only after several months of an ovoviviparous existence. The viviparious fishes are further divided into the aplacentals and the placentals.

Although the presence of organic material in the uterus is universal, the organic content of the egg may be actually greater by 20-40% than the organic content of the animal at birth; in short, the mother has provided protection but probably does not provide organic nutrients to the fetus. At the other extreme, eggs with very little organic material (only 200 mg in *Pteroplatea micrura*) depend almost entirely on the maternal secretions for growth; at birth. *Pteroplatea* contains about 10 g of organic material, an increase of almost 5000%. In contrast, the change in organic substance from egg to newborn pup is only 840 and 1050% in two placental species — *Carcharias glaucus* and *Mustelus laevis*, respectively.

Histoligically, the uterine lining varies from a muscus-secreting layer of cuboidal epithelial cells in species which depend on the egg yolk for nourishment through forms with moderately folded, serous-secreting linings, as in Torpedo, to those with uterine linings beset with *yilli* or *trophonemata* of varying lengths and complexities and glandular surfaces which secrete an abundance of fat (8% of the organic substance in *Trygon violacea*). *Amoroso* (1960) refers to all these uterine secretions as *"uterine milk"* and provides a summary of the composition in various species. *Budker* (1958) describes them as mucous, serous and lipid.

The yolk is mainly digested within the intestine of embryos which depend on it for nourishment. *Te Winkel* (1943) has described the process in the Squalidae. In *Squalus acanthias*, she found that the relatively enormous yolk sac established at an early stage in ontogeny decreased rapidly in size while an internal yolk sec (expansion of the yolk stalk) became relatively larger until the external one remained as only a remnant. Yolk granules are moved by cilia from the external sac through the yolk stalk into the internal sac; then they pass on into the intestine where digestion occurs. All of these structures, including the intestine, are ciliated. Enzymic activity is established in the gut when the embryo is about 60-70 mm long size at birth about 250-300 mm). A small reserve of yolk still remains in the internal sac at birth.

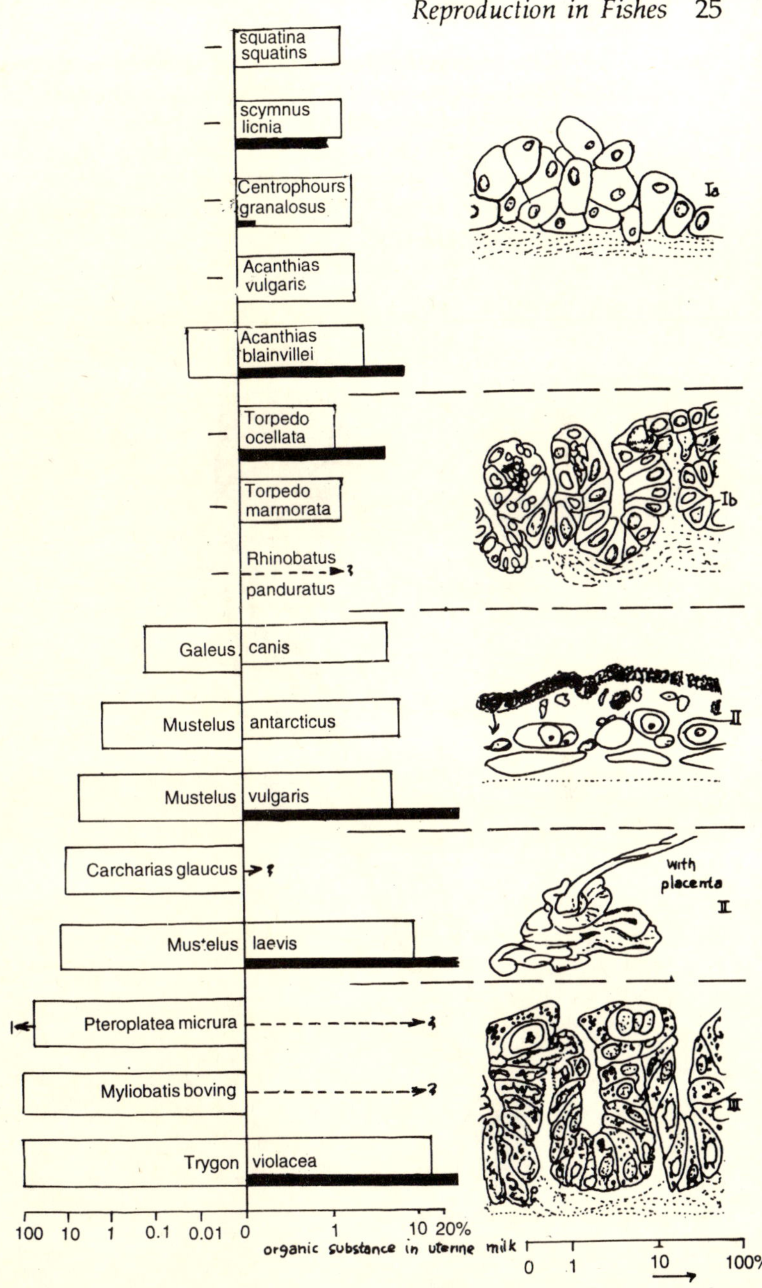

Fig. 1.7. Summary of the maternal-fetal relationships in the elasmobranch fishes. Rectangles to left of ordinate, extent to

which the embryo receives organic substance from mother (negative sign indicates that this does not occur); rectangles to right of ordinate, content of organic substance in uterine fluid (dotted lines show probable values); black bars show degree *to* which maternal liver is reduced during development as measured by factor R. Sketches of the uterine lining are shown at the right.

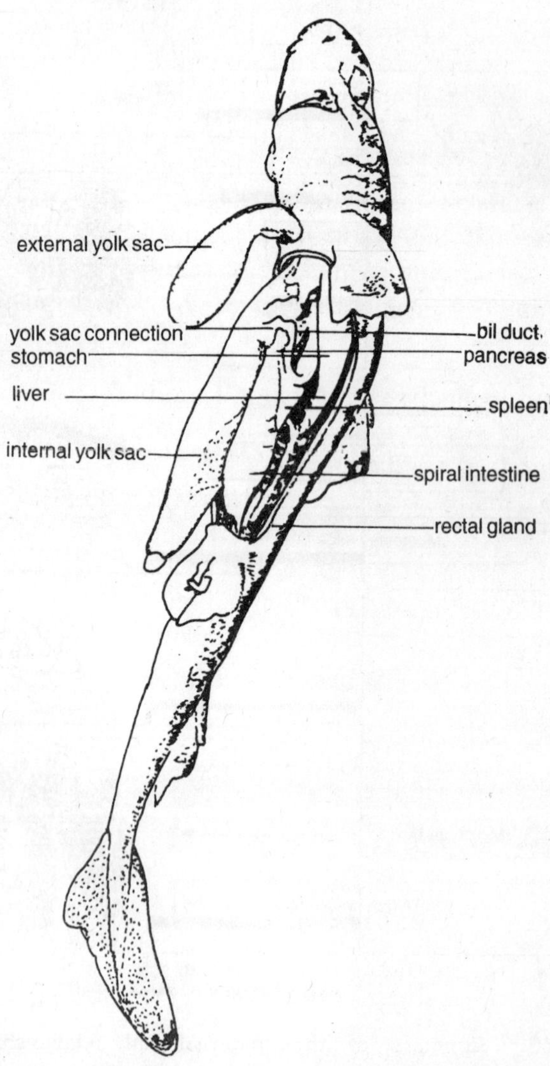

Fig. 1.8.   Embryo of *Squalus suckleyi*

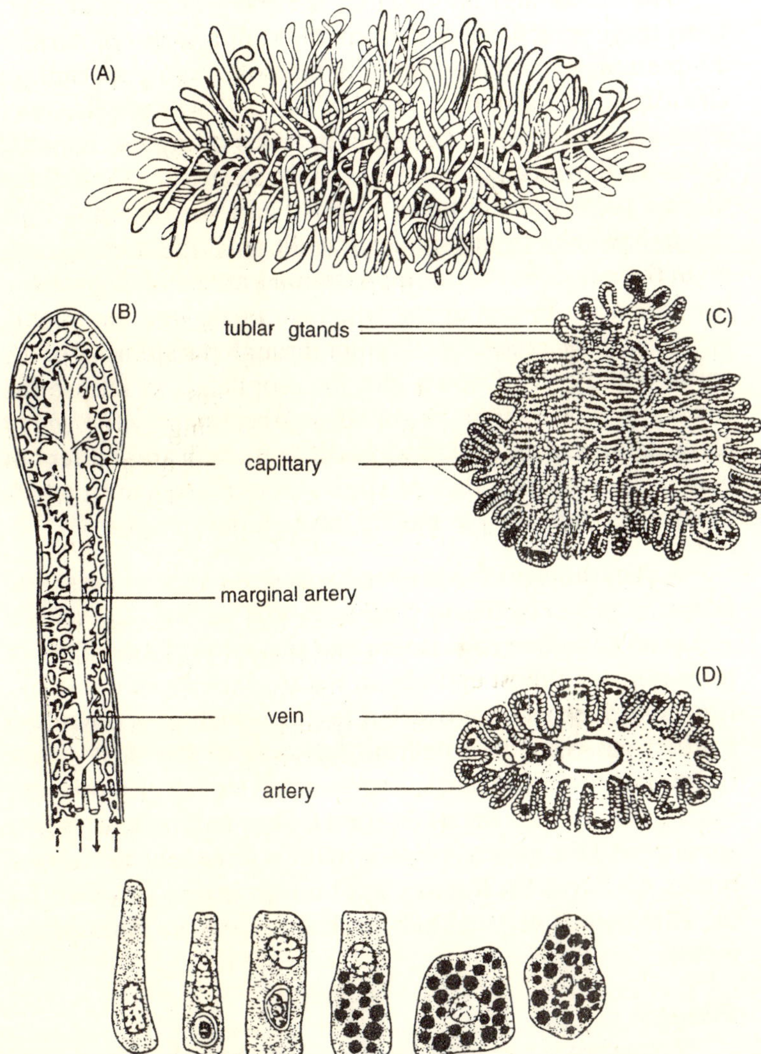

Fig. 1.9. Structures concerned with embryonic nutrition in *Dasyatis violacea (=Trygon violacea)*. (A) Villi of uterine wall, (B) diagram of circulation in uterine villus, and (C) and (D) cross sections of uterine vili (23 g embryo) taken through the apex (C) and base (D) of a villus. Bottom series of cells shows progressive stages in development of uterine gland cells with accumulation of fat droples at right.

The skates and rays display a variety of specializations from those which are oviparous through a series of curious adaptations in ovovivi-parity. In the most highly specialized situations, nutrition by *uterine milk* or *embryotrophe* becomes more efficient than placentation in terms of organic material transferred from mother to fetus. In the electric ray *Torpedo* the uterine secretions are taken directly into the digestive tract through the mouth and spiracles; while yolk digestion is going on in the intestine, the uterine secretions are being digested in the stomach. In one of the stingrays *Pteroplatea*, the highly *glandular trophonemata* or *villi* enter through the spiracles of the embryo and extend down into the esophagus to release the secretions directly into the gut. In another stingray (*Trygon* or *Dasyatis*) and in the eagleray *Myliobatis* the villi are shorter but extremely numerous; they produce a secretion rich in fat which is aspirated into the gut mostly through the spiracles.

A very different-mode of embryonic nutrition in the shark, *Lamna cornubica* by *Shann* (1923). In this animal, egg yolk is absorbed early in development and thereafter the developing pups depend almost entirely on the swallowing of immature eggs and degenerating ovarian tissues which pass down the oviduct packaged in a delicate secretion of the shell gland. Some of these ovoviviparous arrangements are not very different from the oviparity found in the Holocephali, where only a small part of the yolk is enclosed in the yolk sac; the remainder breaks up into a thick milky fluid which is first absorbed by the filamentous external gills and later ingested through the mouth.

Viviparity

Viviparity is common among elasmobranch. The ova of *Mustelus canis*, on arriving in the uterus, become isolated into separate uterine compartments. These are formed by the growth of ridges on the uterine wall which eventually fuse so that each embryo is enclosed in a separate chamber; although uterine compartments are not formed in most of the nonplacental

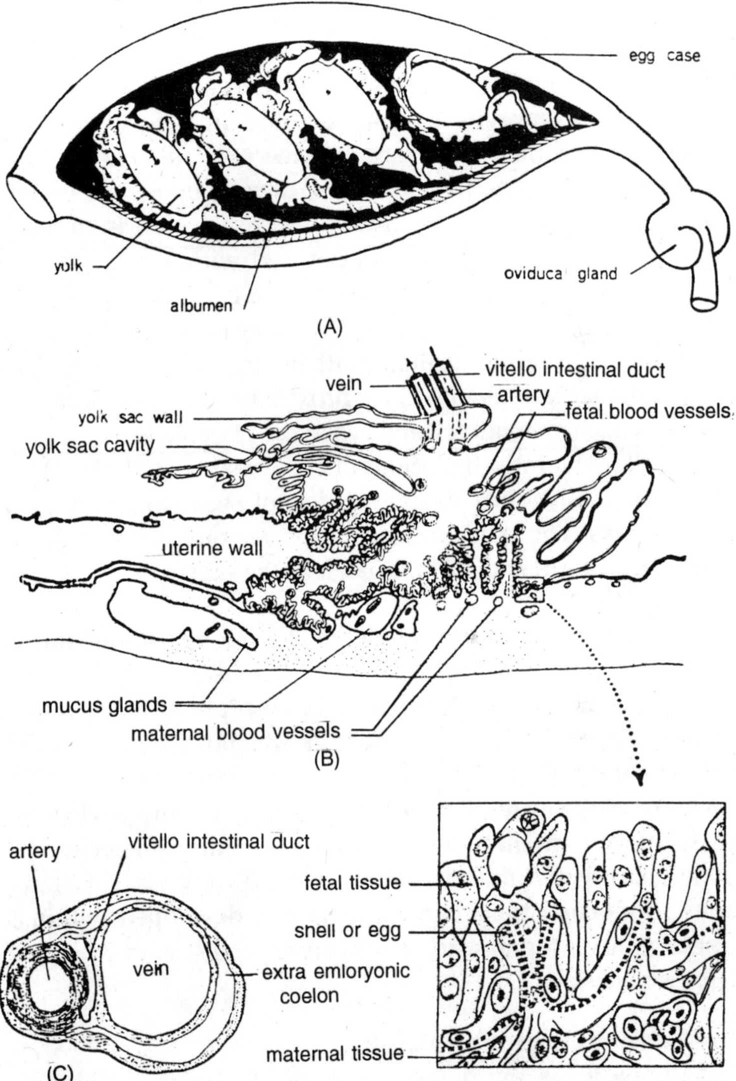

Fig. 1.10. Yolk sac placenta of Mustelus canis (=laevis). (A) Diagram of ventral view of right oviduct and uterus to show orientation of egg cases and embryos, (B) section through placenta with detailed portion at lower right, and (C) transverse section through umbilical cord of 18 g embryo.

sharks there are at least two exceptions (*Galeus canis and Mustelus vulgaris*). *Mustelus canis* leads an ovoviviparous existence during this early period and may not establish a placenta for several months. During this time, while nutrients are obtained from the yolk, these embryos develop elaborate circulatory networks in the walls of their yolk sacs and the yolk sacs themselves become greatly folded. While these developments are taking place in the yolk sac, the uterine mucosa which is initially smooth, loose, edematous, and covered with a columnar or cuboidal epithelium also becomes folded. A placenta is established through the interdigitation of these two series of folds with a thinning of their epithelia to bring the maternal and fetal circulations into close proximity. The *umbilical cord* or stalk which attaches the embryo to the placenta is a modified yolk stalk. The interdigitations characteristic of the placentae of the two fishes just described do not seem to develop in *Carcharhinus*. The specialized yolk sac sits on a modified and extremely vascular discoidal patch of the uterus.

An interesting specialization found in some forms (*Paragaleus, Scoliodon,* and *Sphyrna*) is the presence of numerous villi or *appendiculae* which appear to be absorptive; these suggest that the dependence on uterine milk was not completely lost in the evolution of viviparity among the sharks. In this and in several other features, it is clearly indicated that the phylogeny of viviparity in these fishes was via a highly specialized ovoviviparity and that the steps involved were not very long ones. Both conditions may be markedly developed within a single genus; *Mustelus culgaris is ovoviviparous* while *Mustelus canis* is *viviparous*.

Viviparity Among the Teleosts

Viviparity is the development of the young within the female, occurs in only two orders of teleost fishes, the Cyprinodontiformes and the Perciformes. However, the diversity of mechanisms is as great as that of any other group of vertebrates which range from the oviparous, through several stages of ovoviviparity to highly successful viviparous forms. This, as well as a lack of correlation with particular habitats or geographical regions, argues for a separate evolution on more than one occasion.

It will be noted that internal development in the teleosts always takes place in the ovary, the Mullerian duct, which houses the developing elasmobranch, is absent in the teleosts. Sometimes the young develop within the ovarian follicle to a very advanced stage before ovulation; more frequently, *embryogenesis* occurs within the cavity of the ovary after fertilization within the follicle. Although one might speculate that the ovarian type of gestation is more primitive. C. L. Turner (1947) believes that with one probable exception (*Zoarces*) fertilization precedes ovulation in present-day viviparous teleosts. Bertin (1958c) summarizes the three general situations as follows where "O" represents ovulation, "F" is fertilization, "H" is hatching, and "P" is parturition :

Type Zoarces	O-F		H	P
Type Jenynsia	F	O	H	P
Type Gambusia	F			O-H-P

Both *follicular* and *ovarian gestation* are characterized by ovoviviparous and viviparous types of development. As in the elasmobranchs, the total dry weight of the larva at birth may be less than that of the egg, indicating the complete lack of a maternal contribution to the nourishment of the embryo while in the more specialized viviparous types the egg is endowed with very little yolk and the developing embryo obtains almost all of its nourishment from its parent.

Actually, most of the teleosts which develop within the ovary seem to draw nourishment from the mother and in this sense are not truly *ovoviviparous*. As indicated in figure, only *Sebastes marinus* shows a decline in dry weight between fertilization of the egg and parturition; most of the species

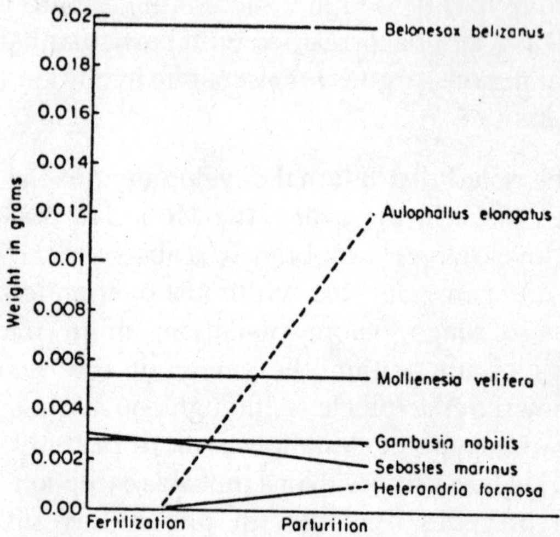

Fig. 1.11. Changes in the dry weight (including yolk) of several viviparous fishes during their development.

show little change in weight indicating that the mother has met the metabolic costs of respiration in the developing tissues so that the total weight of the larva at birth is just balanced by the stored material in the original egg. Only two of the poeciliids (*Heterandria formosa* and *Aulophallus elongatus*) studied by Scrimshaw (1945) and truly *viviparous*, depending on the mother for both their respiratory needs and the anabolic demands for growth. The weight decrease of 34% in *Sebastes marinus* is essentially the same as the value 37% reported for the oviparous trout *Salmo fario*.

Internal Fertilization

Internal fertilization in the teleosts is not necessarily followed by a development of the young in the female. One of the most complex *intromittent organs* described in male teleosts (the *priapium* located on the throat of the Phallostethidae) serves only to fertilize the eggs as they are being laid.

In teleosts, the copulatory organ is usually an enlarged genital papilla or a specialized anal fin. Unlike the

Chondrichthyes, the pelvic fins are rarely used directly in the transfer of sperm; the *priapium* of the phallostethid fishes is again an exception with skeletal elements derived from pelvic fins. An enlarged genital papilla is sometimes referred to as a *"pseudopenis."* Its size bears little relation to viviparity. In the Embiotocidae, a highly successful viviparous group, it appears only seasonally as a modest fleshy papilla; in the Cottidae, an oviparous group, it is often greatly enlarged and may contain erectile tissue. A pseudopenis is found in several other families both oviparous and viviparous. There are many strange modifications of the genital papillae of both sexes. In female *Orthonopias triacis* the protrusible oviduct is smeared with sperms during copulation and then withdrawn to effect fertilization of the eggs which are laid one at a time. In *Apogon imberbis* the urogenital papilla of the female is enlarged and introduced into the male for the reception of sperm. As noted earlier, the elongated genital papilla of *Rhodeus amarus* transfers eggs to the gill spaces of a freshwater mussel. These enlarged genital papillae — both male and female — appear as secondary sex characters during the breeding season, and their developments is probably always regulated by the gonadal steroids.

The copulatory organs of the cyprinodonts are specialized anal fins. The fins are only slightly altered in the Godeidae with a foreshortening of the anterior rays to form a *spermatopodium*. In contrast, the anal fin of the Poeciliidae, referred to as the *"gonopodium"* is profoundly modified through an elongation and specialization of several of its rays; the genital aperture is at its base, but during sexual activity the elongated bony fin rays and the web of tissue connecting these rays interact to form a transitory groove presumably associated with the transfer of sperm bundles to the tip of the gonopodium. The Jenynsiidae and the Anablepidae show a third stage in anal fin specialization to form a large penis containing a permanent tube opening at its tip; at times the organ may be greatly extended; it is a naked structure in the Jenynsiidae but covered with scales in the Anablepidae.

TABLE 1.1

Summary of Modes of Development in Teleosts with Internal Fertilization

Order, family, in species	Type	Fertilization	Gestation	Fetal nourishment and placental structures	Corpora lates
Siluridae *Trachychoristes*	Oviparous	Ovarion cavity or gonaduct	Eggs immediately discharged into water	Yolk in yolk sac and rom aquatic environment	?
Pereiformes Scorpacnidae *Scbastodes*	Oviparous	Follicle zygote retained for long period	Brood remain in ovarion cavity for short period; thereafter development oviparous	Yolk in yolk sac and from aquatie environment	Present
Zoareidue (*Zoarces viviparous*	Viviparous	Ovarian cavity	Retained in ovarian cavity until birth	Very little yolk; ovary furnishes greater part of food; some embryotrophe from dead embryoes; no supplementary structures; absorption through skin	Present
Embiotocidae (*Cymatogaster aggregatus*	Viviparous	Follicle	Hatched into ovarian cavity before segmentation com pleted. Retained for very long period; embryos born in very advanced state. Males sexually mature at birth	Secretions; embryotrophe from desquamated cells; sperm, dead embryos; absorption through hypertrophied yolk sac. Supplementary structures for respiration in form of vascular spatulate processes on fins	Present

Cyprinodontes Poecillidae (*Xiphophorus helleri*)	Oviparous	Follicle	Follicular for entire gestation; high degree of superfetation- embryos very immature at birth; gonads in indifferent stage	Yolk provides most of the intraovarian food, not completely absorbed at birth; "neck" strap forms supplementary structure for absorption of secretions	
Poecillidae (*Heterandria formosa*)	Viviparous	Follicle	Follicular for ent. e gestation; high degree of superfetation	Ovarian secretions; "neck" strap and follicular pseudoplacenta	?
Anablepdae (*Anableps anableps*)	Viviparous	Follicle	Follicular for most of gestation; short time in ovarian cavity	Ovarian secretions; yolk sac hypertrophied and is principal organ of absorption; enlargement of gut forms intestinal pseudoplacenta	Present
Goodeidae (*Ataenobius loweri*) (*Goodea bilineata* (*Zoogonecticus cuitzooensis*) (*Lermicthys multiradiatus*) multiradiatus	Ovoviviparous with superimposed viviparity	Follicle	Ovarian cavity for entire gestation; brood in advanced state of development at birth; high degree of mortality	Ovarian secretions; embryotrophe from stored sperm and dead embryos; Trophotaeniae (fetal) absent in Ataenobius toweri, poorly developed and voluminous in Zoogonecticus cuitzooensis and Lermichthys	Present
Jenynsiidae	Ovoviviparous with superimposed viviparity	Follicle	Ovarian cavity for entire gestation; brood in advanced state of development at birth; high degree of mortality	As for Goodeidae but ovarian fluids imbibed directly by mouth; trophonemata (maternal) in the form of ovarian flaps constitute a branchial pseudoplacenta	Present

Gestation Within the Ovarian Cavity

Sebastodes paucispinis of the family Scorpaenidae *voviviparous* development within the ovarian cavity has been reported on. *Moser* (1967a), who has recently studied this species, found numerous sperms singly or in clumps embedded among the epithelal cells of the maturing follicles or on thier outer surfaces but never within the follicles; he believes that fertilization occurs immediately after ovulation and this would place *Sebastodes* in the "*Zoarces type*" of Bertin's classification. *Sebastodes* embryos lack placentalike connections and depend entirely on their egg yolk for nourishment; they hatch just prior to spawning when the yolk has been largely absorbed and they are about 4-5 mm long; life in the ocean commences as a planktonic larva. Large females may produce broods of as many as two million larvae; the respiratory demands of this large mass of developing tissue must create one of the major problems in maternal physiology. A specialized dual vascular system to the ovary seems to be unique to this group of fishes and is evidently related to these special needs.

The Embiotocidae show an intermediate situation between the strict ovoviviparity of *Sebastodes* and the goodeid or jenynsiid fishes which usually form elaborate placentalike connections. Copulation occurs during the spring or early summer and sperms remain dormant in the ovary until fertilization occurs about 6 months later. Eggs, fertilized within the follicle, are shed into the cavity of the ovary during early segmentation stages and develop there for 10-12 months to an advanced stage (sexual maturity in some males. The *Cymatogaster* egg contains relatively little yolk and depends mainly on the secretions of the highly *glandular ovigerous folds* for nourishment. By the time the larvae have reached a length of 15 mm, spatulate vascular extensions have developed on the vertical fins to assist in nutrition and respiration; these are resorbed just before birth. The alimentary canal begins to function later in gestation and long villi located in the gut are evidently concerned with digestion and absorption of foods entering through the mouth.

Goodeids have more elaborate type of placental connection with progressively diminishing yolk stores. A less significant role for the yolk sac and pericardial sac and an increasing importance of highly vascular rectal trophotaenia of many diverse forms have been reported. These are bathed in the copious secretions of the ovarian cavity; the ovarian secretions are supplemented by degenerating embryos. Similar absorptive surfaces have also been reported in the brotulid fishes.

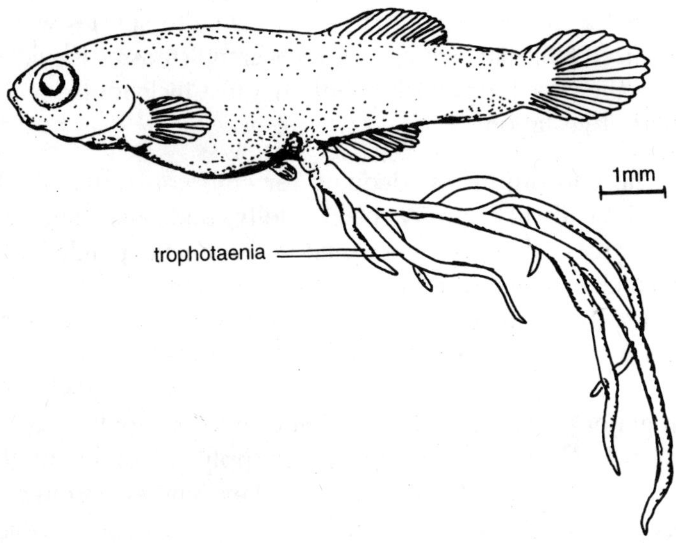

Fig. 1.12.   The trophotaeniae of *Zoogoneticus quitzeoensis* during late gestation

In the jenynsiids there is also an early ovulation following fertilization, an initial dependence on the small supply of yolk, the development of a yolk sac and a pericardial sac. For the major portion of intraovarian development, however, an intimate association is established between the ovarian tissues and the pharyngeal and mouth cavities of the embryo; vascular folds of the ovarian epithelium are in close contact with the gills — an arrangement which *Amoroso* (1960) refers to as a *"potential branchial placenta."* These fishes also imbibe ovarian fluids as a source of nourishment. *Turner* discusses the progressive specialization in ovarian gestation from the strictly ovoviviparous to these several curious placental connections and notes that this trend is evident in a single family — the Goodeidae.

Follicular Gestation

A prolonged follicular gestation has been observed only in the Poeciliidae and the Anablepidae. Again there is a series of forms from the ovoviviparous *(Poecilia)* to species with a specialized follicular *pseudoplacenta;* the complete array is evident within the poeciliids which remain in the follicle for the entire period of gestation.

The yolk supply is adequate for embryonic nutrition in several fishes like guppy, black molly, and swordtail, and respiratory exchanges are effected through an expanded yolk sac which extends over the head as a *"neck strap."* Although it is generally assumed that the intimate association of the portal blood system on the yolk sac and the vascular wall of the ovarian follicles serves only for the exchange of gases and nitrogenous wastes, a transfer of nutrients has not been ruled out. In *Heterandria formosa* the yolk supply is meager and the yolk sac very small; the fetal-maternal exchanges are effected through a greatly expanded and vascular pericardial sac which is wrapped around the anterior part of the embryo. The wall of the follicle is smooth and makes intimate contact with the portal system of the embryo. A further stage of specialization found in this group is an enlarged ventral expansion of the coelomic cavity to form a *"belly sac"* and the formation of finger-like villi on the follicular wall. *Turner* applies the term *"follicular pseudo-placenta"* to this complex of follicular wall, follicular space, and the portal system of the belly sac.

The anablepid fishes develop still another stage in the specialization of the *follicular pseudoplacenta.* Not only is the follicular wall clothed with complex villi but also numerous vascular bulbs appear in the portal circulation of the belly sac.

In addition, there is a spectacular enlargement of the gut in *Anableps*, and this is evidently concerned with the digestion of follicular fluids taken into the alimentary canal an involution of this specialized area of the gut occurs before birth. The enlargement of the gut and the presence of vascular bulbs is unique to the Anablepidae.

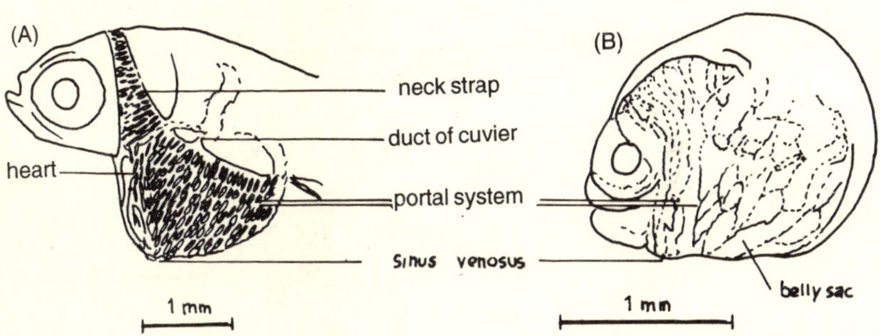

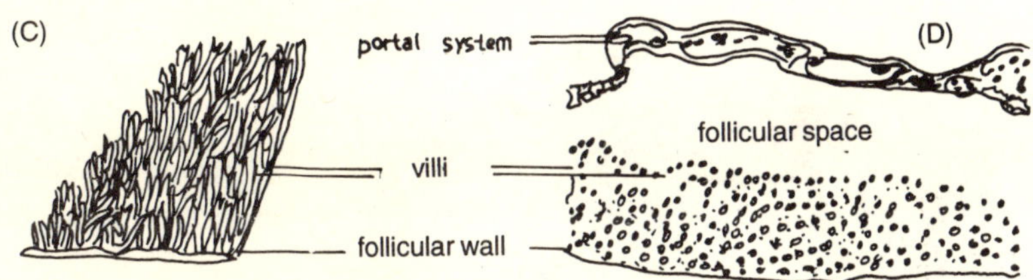

Fig. 1.13. The embryonic-maternal connections in several poeciliids. (A) The portal circulation of the yolk sac of 7.5-mm embryo of Lebistes (=*Poecilia*) reticulatus, (B) Poeciliopsis embryo showing the portal network of the belly sac (pericardial sac anteriorly and expanded coelom posteriorly), (C) low branched villi on the internal surface of the follicle of *Poeciliopsis*, and (D) section of follicle and adjacent Poeciliopsis embryo.

Although the functional morphology has been systematically investigated in the viviparous fishes there are relatively few studies of the physiology. Nutritional demands of elasmobranch embryos were investigated by *Ranzi* (1934) many years ago and there is a small amount of work on their nitrogen metabolism; there are also a few studies of the nutrition of developing teleosts.

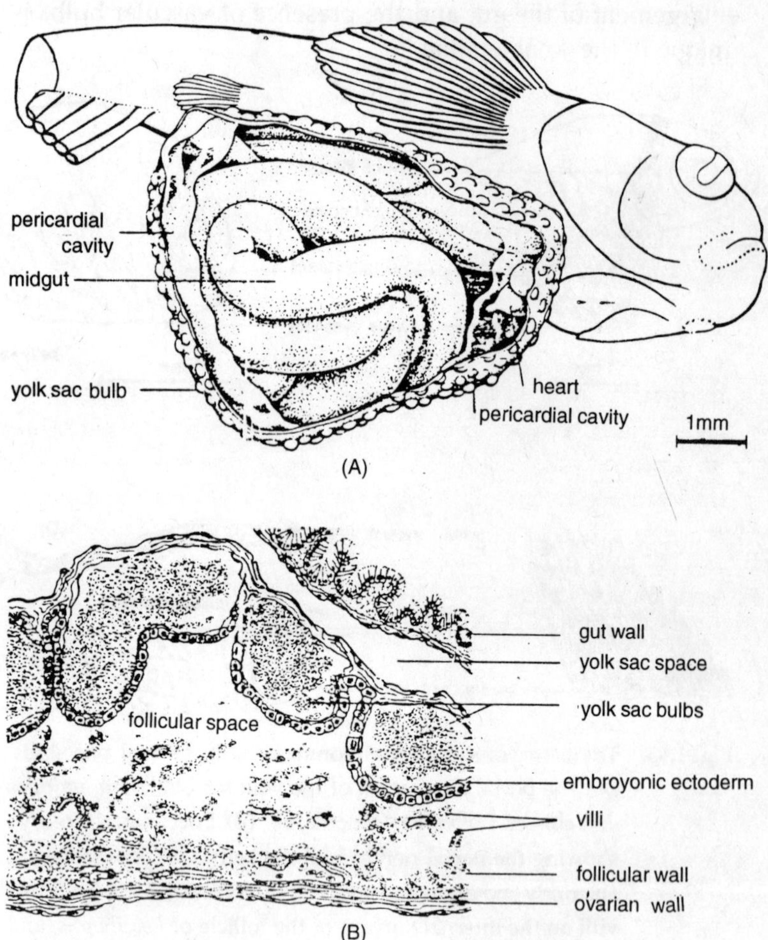

Fig. 1.14. Expanded midgut and maternal-fetal connections in *Anableps anableps*. (A) Dissection of a 21-mm embryo. (B) Section of the wall of a follicle containing a 21-mm embryo.

Superfetation

Superfetation is the development of several broods of young simultaneously within the ovary. The condition has evolved through a shortening of the period required for the development and maturation of the ova; a second group of eggs reaches maturity and is fertilized before the previous group complete development. It must also be related to storage of sperm within the ovary. In many fishes the ovarian epithelium assumes a nurse-cell function and immobilized sperm, embedded in its epithelium, remain viable for many months. *Superfetation* seems to reach a climax in *Heterandria formosa* where live sperm have been found 10 months after a single contact with the male and as many as nine broods may be developing in the ovary simultaneously with birds at intervals of about 10 days.

The Endocrinology of Reproduction

Fishes are seasonal breeders. The timing of reproduction, regulation of the associated morphological changes, the mobilization of energy reserves for gonadal development, and intricate breeding behaviour are very largely dependent on the glands of internal secretion. Most of the endocrine organs are either directly or indirectly involved since this complexity requires profound metabolic adjustments, associated not only with gonadal development but sometimes also with changes in habitat during the breeding season. The roles of various hormones will be dealt under a separate head. In this chapter only the pituitary gonadotropic hormones and the gonadal steroids will be considered.

THE PITUITARY GONADOTROPINS

1. Phylogeny of Gonadotropic Regulation — Studies in the Agnatha

It is a well-known fact that the pituitary gland regulates certain aspects of reproduction in all vertebrate animals. It is equally true that this control is sometimes less precise and embraces fewer elements of reproduction among the fishes and, further, that there may be marked variations among the different groups of fishes.

Lampetra fluviatilis, hypophysectomized in the late autumn and winter, were followed for a period of 5 months and the changes in the gonads carefully described. At the beginning of the experiment, secondary sex characters were absent, the ovaries contained only small eggs, and the testicular ampullae a high percentage of spermatocytes. During the 5-month period the normal and sham-operated animals matured rapidly, secondary sex characters appeared, the ova increased in size, the testicular ampullae became filled with spermatozoa, and spawning took place in early April.   On the contrary, the hypophysectomized animals remained immature in appearance while gonadal development was completely arrested in the females and markedly delayed in the males. *Larsen* (1965, 1969) finds continued relatively slow growth of the ovaries after hypophysectomy but confirms Dodd's findings in the male.

In the gnathostomes, hypophysectomy is followed not only an arrest of ovarian development but also by a general atresia of the follicles *(corpora lutea formation)* while spermatogenesis is completely blocked in the testis. In the lamprey, *spermatogenesis, spermiogenesis*, and *ovarian growth* seem to be autonomous processes; they are retarded but not suppressed in the absence of a pituitary. The endocrine tissues of the gonads are evidently pituitary-regulated since the secondary sex characters which are thought to depend on them fail to develop in the operated animals.

Dodd et al. (1960) and *Larsen's* investigation (1969) have established the indispensable nature of the pituitary for reproduction in the adult *Lampetra fluviatilis*. The Myxinidae have not yet been critically investigated in this connection. These primitive animals should prove particularly interesting and may well increase our understanding of the phylogeny of gonadotropic controls among the vertebrates. On the basis of the lamprey work, it has been suggested that in phylogeny the metabolic controls involving yolk mobilization and gonadal growth preceded those concerned with gametogenesis.

Chemical Inhibition of the Action of Gonadotropins

Due to the technical difficulties of hypophysectomy the investigations of the pituitary functions of fishes have sometimes been hampered. This is true of some of the teleosts which are particularly popular with fish physiologists (*Gasterosteus* and *Cymatogaster*, for example); the mouth and opercular openings may be very small or the skull extremely deep dorsoventrally with much vascular tissue between the roof of the mouth and the brain or the fish may be difficult to handle postoperatively. *Hoar et al.* (1967) and *Weibe* (1968a) have used *Methallibure* as a chemical blocking agent of gonadotropic functions. This agent, is 1-methylallylthiocarbamoyl-2-methylthiocarbamyl hydrazine. It has been carefully tested in the homeotherms and is now finding a use in agricultural practice.

The reactions of four fishes treated with Methallibure either by injection or by addition to the ambient water have been examined. It appears to be particularly effective in *Cymatogaster aggregata* and *Poecilia reticulata*, but much less active in *Gasterosteus aculeatus;* the work with the goldfish, *Carassius auratus*, is preliminary. At present it is not known whether the variable response is because of species difference in reaction to the substance or species differences in the physiological actions of the gonadotropins; it is also possible that the effect varies with the environment — sea water vs. freshwater. It is of interest that some species of domestic animals are proving more responsive than others; the substance has been particularly effective in regulating the farrowing of pigs.

In spite of the variations in response, all species of fishes studied show clear evidence of the blocking of gonadotropic action in a manner comparable to that which followed surgical hypophysectomy. *Spermatogenesis* is either completely suspended or greatly depressed; yolk deposition ceases, and, in the more responsive species, *steroidogenesis* comes to an end. The side effects of the compound appear to be minimal involving a slight stimulation of the pituitary thyrotrophs because of a mild thiourea action and some depression of the somatotrophs. In histochemical studies of *Poecilia, Leatherland* and *Pandey* (1969) have shown that the most likely locus for the block is in the synthesis of the gonadotropins; there was a sharp and significant decrease in both number and size of these cells and no evidence of suppression of secretory activity in the neurosecretory system of the hypothalamus. Ovine LH was found to partially counteract the inhibitory effects of Methallibure in the gonads of *Cymatogaster aggregata.*

The Pituitary-Gonadal Relations in Gnathostome Fishes

In gnathostome fishes like elasmobranchs and teleosts the pituitary-gonadal relations have now been investigated. In the hypophysectomized male dogfish, *Scyliorhinus caniculus. Dodd* and his associates (1960) found that the effects of hypophysectomy were localized in the transitional zone between spermatogonia and spermatocytes. The zonated structure of the dogfish testis was particularly helpful in localizing these effects. It is of considerable interest that the seasonal cycle — presumably regulated by the pituitary — creates an identical picture. Spermatocyte production ceases during the springtime or following hypophysectomy, but in both cases spermatocytes already formed continue to differentiate and become mature spermatozoa. *Dodd* has reached the tentative conclusion that the gonadotropin(s) are essential for the normal transformation which occurs when the spermatogonia become primary spermatocytes. The testicular changes are well marked in dogfishes 6 weeks after hypophysectomy.

Removal of the ventral lobe of the female dogfish pituitary led to cessation of egg laying within 10 days to be followed somewhat more slowly by distinct histological changes in the ovaries. *Vitellogenesis* ceases and all eggs larger than 4 mm become atretic; these changes are evident 3 weeks after hypothysectomy, but it requires about 14 months for the disappearance of all yolk and the reduction of the gonad to the juvenile condition. This *atresia* which followed hypophysectomy

in the absence of the pituitary spermatophores form from the later spermatocytes but they rupture and are resorbed without a discharge of the sperm. Pandey's work (1969c) on the juvenile guppy has also emphasized the interplay of both pituitary gonadotropins and androgens in the maintenance of spermatogenesis.

The changes in ovarian histology following hypophysectomy have been much less frequently examined. However, the findings are consistent with those recorded for the elasmobranchs and indicate that after hypophysectomy a slow but inevitable resorption of ova may be expected; vitellogenesis comes to an end; new ova fail to form; and ovulation and spawning do not occur unless yolk deposition was complete prior to the surgery. *Yamazaki* (1965) found that relatively few goldfish ovulated even though the operation was performed only a few hours prior to the normal ovulation time.

The endocrine-producing tissues of the gonads seem to be pituitary regulated in all of the vertebrates, and regressive changes may be expected in them after hypophysectomy. Atrophy of the testicular interstitial tissues has now been recorded in several teleost fishes and in lobule boundary cells. Regressive changes in the *Sertoli cells* have also been noted. The gonadotropic inhibitor *Methallibure* causes atrophy of both the interstitial cells and the *Sertoli cells* in the surfperch *Cymatogaster*; both of these groups of cells are actively concerned with steroidogenesis in this species. All these data are consistent with the concept of a pituitary control of steroidogenesis in the gonadal endocrine tissues. Regressive changes noted in the efferent duct system are probably secondary to the degeneration of the steroid-producing tissues.

Effects of Hypophysectomy on Gestation

The association of cytological cycles in pituitary secretory activity with the pregnancy cycle of viviparous fishes have been done by seasonal workers. Ranzi (1936, 1937) examined the endocrne glands in a part of his extensive studied of selachian viviparity and described pituitary hypertrophy and hyperemia with a decreased acidophilia in pregnant females. Pitutiary changes have also been assciated with the reproductive cycles of viviparous teleosts. However, in both groups of fishes these pituitary cycles may only regulate the timing of reproduction and the development of the gametes as they do in the oviparous species. From present studies it must be concluded that the pituitary plays little or no part in the maintenance of pregnancy in the elasmobranchs but may be involved in some teleosts.

In the late thirties, *Hisaw and Abramowitz* (1938, 1939) reported that pregnancy was uninterrapted by hypophysectomy in the viviparous dogfish, *Mustelus canis*; animals hypophysectomized in early stages of pregnancy showed normal development of young for a period of $3\text{-}^1/_2$ months. These seem to be the only experiments of this nature recorded for the elasmobranchs, although *Dodd et al.* (1960) report that removal of the ventral lobe of the pituitary of *Scyliorhinus* has no effect on the oviducts which retain their normal secretory capacities.

Studies of the effects of hypophysectomy on gestation in the viviparous fishes are likewise preliminary. *Ball* (1962) found that 12 *Mollienesia latipinna*, hypoph· sectomized during pregnancy, gave birth to normal offspring, the embryos completed their development in the usual time and were as lively and survived as well as the controls. *Chambolle* (1964), however, working with another viviparous cyrinodont *Gambusia*, recorded a highly significant mortality of embryos when females were hypophysectomized during the first days of pregnancy; in part, the mortality was thought to be related to a disturbance in water and electrolyte metabolism — perhaps associated with the loss of the pituitary — adrenal control. Thus, the evidence from two different species of viviparous cyprinodonts is contradictory but there are further indications from other studies that the pituitary is essential to successful gestation and parturition in some teleosts. *Wiebe* (1967) found that treatment with the gonadotropin-blocking agent *Methallibure* leads to an atrophy of the secretory ovigerous epithelium in *Cymatogaster*. Associated with this, *Wiebe* noted reduced growth and a 39% mortality of the intraovarian embryos. These results are preliminary and based on only three pregnant females; the effects may also be indirect ones mediated through the gonadal steroids. In addition, there are several reports of premature release of young by pregnant cyprinodonts following injection of mammalian pituitary preparations. It is also pertinent that incubation and parturition of developing young in the male sea horse. *Hippocampus hippocampus*, is disturbed by hypophysectomy although gonadectomy is without effect. Finally, there is some suggestive evidence of a role for prolactin in the gestation of certain cyprinodonts and an intimation that the neurohypophysial hormones are active in parturition.

Biochemical Nature of the Gonadotropin(s)

Two separate gonadotropins, the *follicle-stimulating hormone* (FSH) and the *luteinizing hormone* (LH), are physiologically distinct in all tetrapods. Although the nature of the fish gonadotropins is still not resolved, the data for a single factor seem to be accumulating steadily. In addition to the physiological literature summarized below, the biochemical evidence for a single protein increases as the preparations have been purified more carefully. Most of the studies — both physiological and biochemical — have been concerned with teleosts.

When fish pituitary extracts are tested in tetrapods, both the FSH- and the LH-like effects are frequently elicited. *Witschi* (1955) assayed pituitaries of sharks, garpikes and salmon — representatives of the three major groups of jawed fishes. Positive evidence for FSH was obtained by using the vaginal cornification test in the rat and for LH by using the weaver-finch feather reaction. The FSH content was very low; most of the gonadotropic activity appeared to be associated with LH. *Ball* (1960) has summarized these and other investigation —

most of which provide suggestive evidence for the presence of the FSH-like effect and strong evidence for the LH action when fish pituitaries are injected into mammals. A series of papers by *Otuska* (1956a, b, c, 1957) using partially purified salmon pituitaries injected into newts and mice are in agreement with evidence for two gonadotropic proteins with distinct physiological effects. However, not all of this evidence for an action of fish gonadotropin(s) in higher vertebrates is positive. *Burzawa-Gerard* and *Fontaine* (1965) prepared a carefully purified extract of carp pituitaries which appeared to be a single protein and was entirely inactive in mammals.

Evidence concerning the nature of fishes gonadotropin has also been sought in tests with the purified mammalian fractions injected into fishes. The preparations of mammalian FSH have consistently given negative results. LH, on the contrary, usually elicits both the gametogenetic and the steroidogenetic actions commonly associated with two different fractions in the higher vertebrates. Among the more recent studies are those of *Sundararaj* and *Goswami* (1966) who report that LH alone, of several pituitary factors tested, would stimulate ovulation in *Heteropneustes; Ahsan* (1966b) who found a partially purified salmon gonadotropin to have an almost identical physiological action to mammalian LH when injected into hypophysectomized *Couesius plumbeus: Ramaswami* (1962) who has used enzymes which selectively digest or inactivate the gonadotropic factors in mammalian pituitaries (trypsin or pepsin for LH and ptyalin for FSH) and found convincing evidence for the LH-like factor but no support-for FSH in *Heteropneustes fossilis. Yamamoto* and *Yamazaki* (1967) for the acceleration of ovulation in hypophysectomized goldfish. However, it cannot yet be concluded that the fish gonadotropin is always physiologically similar to mammalian LH. By using the spermiation test in goldfish, *Yamazaki* and *Donaldson* (1968a) have obtained strong reactions with a partially purified salmon gonadotropin and with human chorionic gonadotropin (HCG) but no reaction with either FSH or LH.

The mammalian chorionic gonadotropins are often used for experimental purposes and in fish culture work because of their ready availability.   These factors, produced by the mammalian placenta during later stages of pregnancy, mimic many of the activities of the pituitary gonadotropins although they are definitely known to be different biochemically from the pituitary factors.   Two chorionic gonadotropins are in common use: one, prepared from the serum of pregnant mares (PMS), usually has biological properties similar to a combination of FSH and LH; the other, prepared from human pregnancy urine (HCG), acts more like LH.   In mammals, the precise effects depend somewhat on the dosage used.   When injected into fish, HCG has often been found to be quite effective and this is in line with its LH-like activity.   Pregnant mare serum is sometimes active but has usually been less consistently effective than HCG.   *Sundararaj* and *Goswami* (1966) report that 100 IU/ fish of HCG will induce ovulation in *Heteropneustes* while 250 IU/fish of PMS are required to elicit the same reaction.

Both biochemical and physiological evidence, now indicate a single proteinaceous gonadotropic factor in the pituitaries of teleost fishes.   Although this may sometimes mimic FSH and frequently mimics LH when injected into tetrapods, it is clearly not identical with either of these factors.   It more closely resembles LH but has some unique physiological properties and may be expected to differ from its counter-part in the tetrapods and even differ in its effects in different groups of fishes. Investigations of some of the more primitive groups of fish should be rewarding.   The lactogenic hormone (prolactin, luteotropin) is also gonadotropic in some mammals (the ovary of the rat), but no evidence has yet been presented for a comparable action in the lower vertebrates.

THE GONADAL STEROIDS

The development of the secondary sex characters prior to breeding has been shown to depend on gonadal steroids in the majority of fishes. Following gonadectomy, secondary sex characters fail to develop or, if seasonal in occurrence, regress. Their differentiation is initiated or stimulated by a wide variety of androgens and estrogens — both natural and synthetic. This may be the only broad generalization possible; in all the later events associated with reproduction like ovulation, spermiation, spawning, and breeding behaviour, the division of regulatory responsibilities between pituitary and gonads appears to be rather variable in different species.

The physiologically most active gonadal hormones of higher vertebrates are *testosterone* from the interstitial cells of the testis, *estradiol-17β* and its derivatives from the ovarian follicle, and *progesterone* from the corpus luteum. The biogenesis of all these compounds is from acetate via cholesterol, as the parent sterol, to *testosterone* through intermediate steps involving *progesterone;* the *estrogens* are derivatives of *testosterone* or closely related molecules and form the terminal part of the biosynthetic chain. Thus, the biologically active gonadal steroids are linked through common pathways and one may expect to find small amounts of any or all of them in tissues which are primarily concerned with the synthesis of a single important hormone. The biogenesis of the *adrenocortical steroids* is linked to the same chain.

These metabolic pathways are very ancient phylogenetically. *Progesterone, estradiol-17β,* and some other *estrogens* have been identified in the ovaries of invertebrates including echinoderms and mollusks as well as the lower vertebrates. *Estrogens* are also well-known in plant tissues. It is not then surprising that both *androgens* and *estrogens* of several biochemical sorts have been regularly found in the gonadal tissues and blood of many different fishes. However, the presence of a gonadal steroid is no evidence of its significance as a hormone and the much more difficult task is to identify steroids which are physiologically important in regulating the reproductive activities of fishes. At present, it is by no means certain that all the steroids recognized as physiologically active in higher vertebrates are biologically significant in the fishes. Different groups of fishes seem to react somewhat differently to some of the same steroids. Moreover, it seems certain that there are important biosynthetic pathways of steroidogenesis in fishes which have not been recognized in the higher vertebrates. In fishes, as in the higher vertebrates, some of the synthetic compounds are more active than the naturally occurring ones.

The Androgens

There is good evidence that pathways recognized for biosynthesis of androgens in higher forms are operating in the elasmobranchs and that *testosterone* is an important end product in the testis. *Chieffi* and *Lupo* (1961), using chemical methods, investigated the testicular tissues of mature dogfish, *Scylliorhinus stellaris*, and recorded *progesterone* (100 µg/kg), *testosterone* (50 µg/kg), *androstenedione* (70 µg/kg), and *estradiol-17B* (20 µg/kg). *Idler* and *Truscott* (1966) were the first to isolate *testosterone* from the blood of a male elasmobranch. They found values which were relatively high (7.4 µg/100 ml in *Raja radiata*) when compared with human males (about 0.56 µg/100 ml). *Testosterone* was also found in the blood of female *R radiata*, but the mean values were much lower. *Progesterone, androstenedione, androsterone,* and other steroids have been isolated from the semen of the dogfish, *Squalus acanthias* and *testosterone* biosynthesis was demonstrated by incubation techniques in the testis of this fish.

A number of different *androgens* have been isolated from the tissues and blood of teleost fishes. Steroid values have also been recorded for testicular tissues of the teleosts *Morene labrax, Mugil cephalus* and *Serranus scriba* the latter is a hermaphroditic form.

Testosterone has been found in significant amounts together with several other steroids such as *17x-hydroxyprogesterone*, recognized in higher forms as a step in testosterone biosynthesis and *androsterone* — a product of testosterone synthesis. As might be expected from the known relationships of these substances, androgens have also been found in the tissue of female fish. *Testosterone* values are extremely variable (ranging up to 17 μg/100 ml of male blood); since tissue values must represent a balance between synthesis and utilization it is always difficult to relate the actual amounts to the physiology. Three particular facets of Idler's studies require special comment: (a) the presence of *11-ketotestosterone*, (b) the occurrence of conjugated testosterone, and (c) the marked variations associated with the life cycle of the salmon.

Idler et al. (1960, 1961a, b) were the first to identify *11-ketotestosterone* as a natural product and subsequently to show its biological activity as an androgen in stimulating the development of secondary sex characters in salmon and chickens. The isolation of *11-ketotestosterone* from both males and females of Atlantic and Pacific salmon suggests that the synthetic pathways of the *androgens* in these fishes are somewhat different from the usually recognized ones. This steroid is present in amounts up to 17 μg/100 ml of blood and occurs along with several of the more familiar androgenic compounds including *testosterone*. *Idler* and *Truscott* (1963) found that testosterone and 17-hydroxyprogesterone will serve as precursors for *11-ketotestosterone* in the sockeye salmon and, subsequently *Idler* and *MacNab* (1967) demonstrated its *in vitro* synthesis from *adrenosterone* and *testosterone* in Atlantic salmon gonads and sperm; they suggest the probable pathways of biosynthesis. *Arai* and *Tamaoki* (1967) also studied the *in vitro* synthesis in Atlantic salmon gonads and sperm from *adrenosterone* and *testosterone*; they too suggest probable pathways of biosynthesis. *Arai* and *Tamaoki* (1967) have studied the *in vitro* biosynthesis of *11-ketotestosterone* in rainbow trout, *Salmo gairdneri*.

Conjugated testosterone was first reported in fish blood by *Grajcer* and *Idler* (1961, 1963). In higher vertebrates, the steroids are transported, at least in part, as conjugates with serum proteins and glucuronic acid. *Grajcer* and *Idler* obtained the release of *testosterone* in amounts of 13.7 μg/100 ml of male sockeye salmon blood after treatment with the enzyme B-glucuronidase. The *"free" testosterone* in these samples of blood was 1.7 µg/100 ml. The blood of female salmon also contained conjugated as well as *"free" testosterone* with relatively less of the conjugated form (7.6 µg of conjugated to 7.8 µg per 100 ml of free) even though the total amounts were about the same. Since these initial studies of testosterone glucuronoside, *Idler* and *Truscott* (1966) have reported on conjugated *testosterone* in the skates. *Raja radiata* and *R. ocellata:* these conjugates are probably as common in fish blood as they apparently are in the higher vertebrates.

The quantitative changes in several plasma steroids including *11-ketotestosterone* and *testosterone* have also been observed during the migration of the sockeye salmon. Definite differences were recorded and the shift in ratio of *11-ketotestosterone* and *testosterone* may be significant in the physiology and behaviour of the spawning migration. Changing levels of *androgens* have also been reported in females of the ovoviviparous elasmobranch, *Torpeda marmorata*, during the reproductive cycle. There was a moderate increase in *androsterone* at the end of the gestation period and a marked steady rise in *dehydroxyepiandrosterone* from pregestation to the midgestation period. These variations indicate physiological changes in the demands for *estrogens* rather than imply a role for *androgen* in the gestating ray. In the stickleback, *Gasterosteus aculeatus*, where the development of sexual behaviour is cyclical and associated with strong agonistic behaviour in the males. As reported by *Gottfried* and *van Mullen* (1967) the dominant males have testicular androgen levels which are five to seven times higher than the nondominant individuals; *testosterone* could not be detected in the testes of the nondominant fish but was present in the dominants. These cyclical and seasonal changes in gonadal steroido-genesis are presumably triggered by variations in the gonadotropic activity of the pituitary; seasonal changes in the latter have been traced in both place *Pleuronectes* and the perch Perca.

The Estrogens and Progesterone

At all levels in phylogeny *Estradiol-17β* has been found in the ovaries of many fishes the lampreys, *Petromyzon marinus* dogfishes, *Squalus suckleyi* and *Scyliorhinus caniculus* the ray, *Torpedo marmorata*, the ratfish, *Hydrolagus colliei* the lungfish, *Protopterus annectes* and many different teleosts. This wide distribution — together with its presence in several invertebrate phyla — suggests that it will be found in at least small amounts wherever there is active steroid synthesis. *Estradiol-17β* has also been identified in the blood of elasmobranchs and teleosts and is probably the physiologically most important estrogen. *Gottfried* (1964) tabulates *estradiol-17β* values for ovarian tissues ranging from a trace to 126 µg/kg. with the most usual amounts varying from 10 to 20 µg/kg.

Estradiol-17β is recognized as the parent substance for several other *estrogens* known to occur widely in the animal world. Particularly frequent in occurrence are *estrone* (an oxidation product of estradiol-17β) and *estriol* which is derived from *estrone* by 16-hydroxylation and reduction at 17. Both *estrone* and *estriol* were found in the ovaries of many of the species, but the proportions are quite variable and occasionally one of the compounds is absent. *Cedard et al.* (1961) traced seasonal changes in the estrogens of the blood of both male and female Atlantic salmon. The total *estrogen* content in both sexes showed a five- to sixfold increase at the time of spawning, reaching 5-6 µg/100 ml of blood. *Estrone* was found at all seasons; *estriol* appeared in significant amounts only at spawning; and *estradiol* was only present in small amounts and seemed to disappear completely in the females at spawning and in the postspawning males. Again, it is difficult to interpret the findings in terms of physiological demands but results such as these emphasize the hazards of conclusions based on a few estimations of the gonadal steroids at only one season.

Progesterone has been found in the tissues of all vertebrates and many of the invertebrates. It is probably ubiquitous as a link in steroid biogenesis. As yet, however, there is no definitive evidence that *progesterone* is a physiologically active hormone with distinct endocrine responsibilities. *Hisaw* (1959, 1963) concludes the role of progesterone as a hormone.

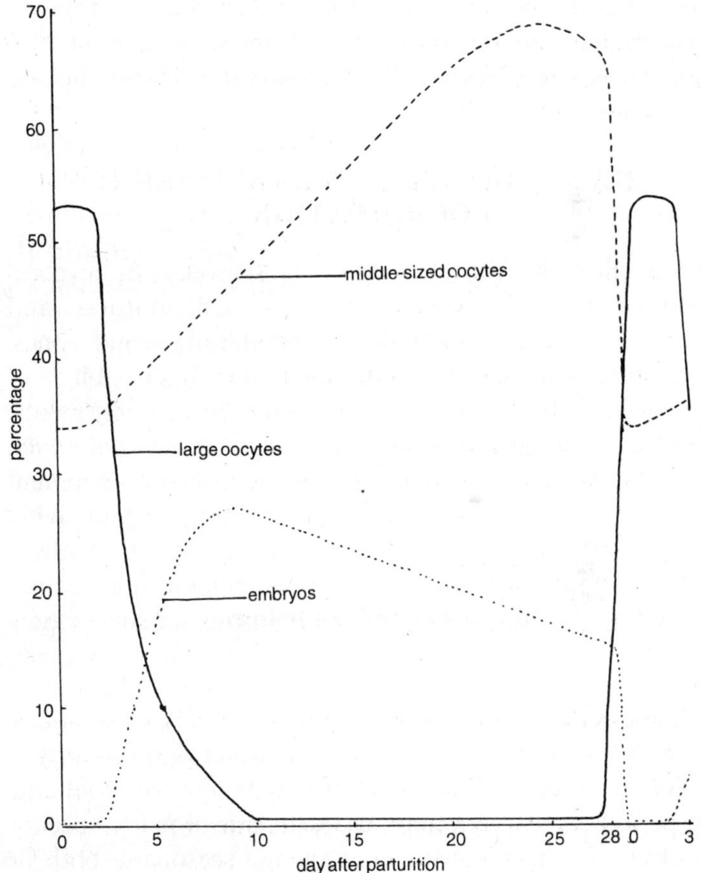

Fig. 1.16. Variations in the number of oocytes and embryos during the 28-day gestation period of *Poecilia*. The number was calculated as a percentage of the total number of structures present in each ovary per 24 hr.

Chieffi has convincingly the development of corpora lutea with the stages of gestation in the electric ray *Torpedo*. He has also demonstrated histochemically the biosynthesis of a variety of steroids — including *progesterone* — in the corpora lutea, while chemical methods have identified *progesterone* in *Torpedo* blood during pregestation but again the question of its presence as hormone or precursor has not been settled. It appears, however, that active *steroidogenesis* occurs in the corpora lutea of some fishes; it seems equally evident that this is not the case in all fishes

REPRODUCTIVE CYCLES AND THEIR COORDINATION

Reproduction is almost always a seasonal or cyclical phenomenon. An annual cycle of temperatures and photoperiods characterizes the temperate and frigid zones; the rainy seasons may markedly alter freshwater habitats of the tropics. In these seasonally unstable environments, reproduction is geared to take advantage of seasons which offer the greatest opportunities for survival and development of the new generation. Even where conditions are relatively stable and the eggs or young are produced regularly throughout the year, there may still be a cycle of gonadal maturation imposed by the energy demands of maturing a batch of eggs or young.

These cycles of gonadal development frequently alter many aspects of metabolism as well as behaviour and reproductive physiology. Cycles of active feeding with storage of fat and long periods of starvation are characteristic of many species. The electrolyte metabolism may change seasonally both in species which inhabit the ocean waters of relatively constant salinity and in the *euryhaline* and *anadromous* forms. The basic cycle is probably the one imposed by the seasonal nature of reproduction, and these regular changes in metabolism are secondary to it.

The endocrine system forms the major link between the environment and the organs concerned with reproduction. Changing environmental conditions, operating through the sensory system and specific centers in the brain, trigger neurosecretions which in turn regulate the activities of the pituitary gland. The pituitary hormones have direct effects on *gametogenesis*, metabolism, and behaviour. These hormones also regulate the development of the gonadal endocrine tissues. Gonadal hormones take over some of these pituitary responsibilities and carry on the coordination of events in the production of gametes, sexual behaviour, fertilization, and sometimes parental care. There are, in fact, several cycles within the gonads, but many of the details of regulation at the cellular level remain to be unraveled. Studies of the cyclical changes in the structure of the gonads and endocrine organs, as well as variations in actual secretion of hormones, form one of the most voluminous components of the literature concerned with reproduction.

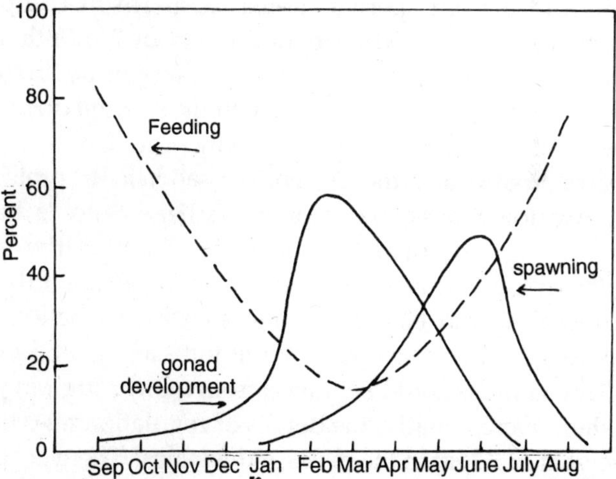

Fig. 1.17. Relation between *feeding* cycle (percent sample with food in stomach) and reproductive cycle (*gonad development*, percent fish with ripening gonads; spawning, percent ripe (fish) of the haddock, Melanogrammus aeglifinus.

A definite breeding season is absent in the female of the spotted dogfish, *Scyliorhinus caniculus,* and in the male

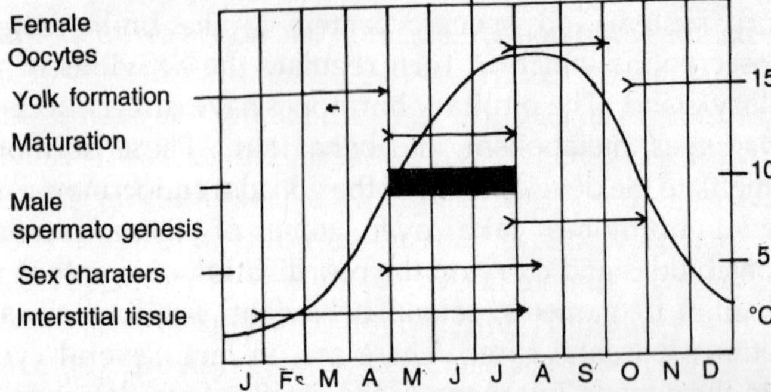

Fig. 1.18.   Seasonal cycle of reproduction in Gasterosteus. Ordinate, months: curve, temperature in °C; horizontal black bar, spawning period; horizontal arrows, periods when more than 50% of samples showed characters listed on the left.

spermatogenesis is continuous with a regular progression in the ampullae from spermatogonia to sperm in comparison there is a limited breeding period of about 2 months in the spiny dogfish, *squalus acanihias,* and during this period there is a maximum accumulation of semen in the efferent ducts which is followed by a pause in sperm production. This suspension of sperm production is the result of a cessation in the proliferation of spermatogonia several months earlier. Since a band of degenerating ampullae appears in the testis of the spotted dogfish following hypophysectomy, there seems little doubt that these cyclical changes in the sperm production of the spiny dogfish are regulated by the pituitary gonadotropins. Similar examples could be drawn from many different groups of fishes. Occasionally, the details of regulation may even be somewhat different in the males and females. There are diurnal as well as seasonal cycles. In the cyprinodont, *Oryzias latipes,* there is a daily cycle of ovlution (between 1 a.m. to 5 a.m.), mating behaviour (4 a.m. to 7 a.m.), and the laying of fertilized eggs which follows soon after mating. Ovluation and oviposition are indepedent phenomena; the former depends on the temperature and the light cycle while the latter depends on contact stimuli associated with the sexual embrace. In another cyprinodont, *Rivulus marmoratus,* there is an internal self-fertilization which is also timed by the daily light cycle. *Rivulus marmoratus* is hermaphroditic with functional ovotestes. The timing of events within these organs is so correlated that there is a peak frequency in ovulation; fertilization occurs at dawn with a peak in oviposition at noon. *Harrington* (1963) also found evidence of seasonal changes in the cycles in accordance with light conditions.

2

Variety in Development

Over the past few years improvements in technique for rearing marine fish have increased the number of species available for experiment and for *aquaculture*. These techniques are summarized by *Kinne* (1977), *Blaxter* (1981), *Hunter* (1984). The greatest advance has been the use of very small food items, especially of the rotifer *Brachionus plicatilis*, but also the naked dinoflagellate *Gymnodinium splendens* and other organisms such as *Mytilus* trochophores and sieved natural zooplankton, as a food source in the very young stages when the size of the mouth is limiting. Interest is now increasing in the use of *"green-water"* culture, where the larvae are maintained in fairly high densities of algae such as *Chlorella*, which may damp out metabolite fluctuations, perhaps improve oxygenation, and provide a secondary food source for the larvae. In the future the use of compounded diets, especially in the form of microcapsules small enough to be eaten whole, may provide a further breakthrough. *Appelbaum* (1985) reared Dover sole *Solea solea* entirely on compounded diets and cited other similar successful work on plaice *Pleuronectes platessa*, vendace *Coregonus albula*, sea bass *Dicentrarchus labrax*, turbot, catfish *Clarias gariepinus*, and the Atlantic silverside *Menidia menidia*. The best survival rate of sole on an artificial diet was obtained when live brine shrimp *Artemia nauplii* were provided for the first 10 days of feeding. *Artemia* is certainly still the staple live food, both in experimental and applied fish culture. It has become increasingly obvious that *Artemia* from different sources can vary in quality — for example, in fatty acid *"profile"* —

and success and failure in the past may have hinged on this hitherto unappreciated factor. Work on some of the nutritional aspects of rearing fish larvae and the relevance of digestive processes was done by *Dobrowski* (1984).

Other factors in rearing, such as optimum food density, stocking density, type of tank, light, and other environmental conditions, have now been established. Production of eggs out of season by the use of artificial photoperiods, temperatures, and homone injections have also greatly improved the availability of larvae year-round.

One of the most striking advances has been the improvement in survival and growth when larvae are reared in the absence of predators in large-scale facilities, or *"mesocosms"*, in the form of large onshore tanks, large plastic-walled cylinders sited in sheltered coastal waters, or impounded coastal bays or lagoons. Atlantic cod *Gadus morhua*, turbot, and red drum *Sciaenops ocellatus* have been reared with unprecedented success.

Progress and Diversity of Development

The early life history stages of fishes and their characteristics are discussed more generally by *Blaxter* (1969), *Hempel* (1979), and *Kendall et al.* (1984). Great variety exists from species to species and, in particular, the size and extent of differentiation when the young fish first becomes free-living is of considerable significance for its chance of survival. During the final ovarian maturation of the eggs of marine fish such as the Atlantic cod, whiting *Merlangius merlangus*, haddock *Melanogrammus aeglefinus*, and plaice, there is a massive uptake of water and concomitant reduction in protein phosphate. This influx of water, such that the water content may reach as high as 92% of the egg weight, is an adaptation to pelagic life because the egg fluids are hypotonic and make the eggs buoyant. Freshwater fishes with demersal eggs, like the rainbow trout *Salmo gairdneri*, powan *Coregonus lavarcticus*, and pike *Esox lucius*, do not show these changes. The initial buoyancy of pelagic fish eggs, like those of the flounder *P. flesus*, depends on the salinity in which the female is kept before spawning. Females from low salinities tend to produce eggs that are neutrally buoyant at lower salinities, buoyancy, of course, not only being affected by water content but also by the osmolarity of the egg fluid.

There is great variety in the reproductive styles of fishes. In most of species, the eggs develop independently, but there are many instances of parental care. In littoral species such as cottids, blennies, and gobies, this takes the form of guarding the eggs, but nests may be built with one or other parent guarding and often ventilating the nest. Mouth brooding of eggs and larvae is found in cichlids such as tilapia and in ariid catfish. Other species have evolved ovoviviparity or viviparity, the former where the eggs develop within the female, the latter where nourishment is provided via *"pacental"* structures (*trophotaenia*) within the female. While care is more common by the male parent care (in 61 families) by the female occurs in 41 families. Care by the male is clearly linked to the prevalence of external fertilization in fish and generally to *polygamy* and male territoriality.

Typically marine eggs are single, buoyant, and with a modal diameter of about 1 mm (although the range is from 0.6 to 4.0 mm). Most freshwater fish lay demersal eggs with a modal diameter somewhat greater than 1 mm. The eggs may merely rest on the sub—stratum, or have some means of attachment such as adhesive threads or a supporting pedestal. In species such salmonids the eggs are buried in the gravel, and the grunion *Leuresthes tenuis* lays its eggs intertidally in the sand. Other types of demersal eggs are found in some littoral marine species, and, in the more offshore Atlantic herring, capelin *Mallotus villosus* and Pacific cod *Gadus macrocephalus*.

TABLE 2.1

Classification of Reproductive behaviours

A.   Nonguarders

   1.   Open and substratum spawners

      a.   Pelagic spawners
      b.   Rock and gravel spawners with pelagic larvae
      c.   Rock and gravel spawners with benthic larvae
      d.   Nonobligatory plant spawners
      e.   Obligratory plant spawners
      f.   Sand spawners
      g.   Terrestrial spawners, in damp conditions

   2.   Brood hiders

      a.   Beach spawners; above waterline at high tides
      b.   Annual spawners; eggs estivate
      c.   Rock and gravel spawners
      d.   Cave spawners
      e.   Spawners in live invertebrates

B.   Guarders

   1.   Substratum spawners

      a.   Pelagic spanwers; at surface of hypoxic waters
      b.   Above-water spawners; male splashes clutch
      c.   Rock spawners
      d.   Plant spawners

   2.   Nest spawners

      a.   Froth nesters
      b.   Miscellaneous substratum and materials nesters
      c.   Rock and gravel nesters
      d.   Glue-making nesters
      e.   Plant material nesters
      f.   Sand nesters
      g.   Hole nesters
      h.   Anemone nesters: at base of host

C.   Bearers

   1.   External bearers

      a.   Transfer brooders; eggs carried before deposition
      b.   Auxiliary brroders; adhesive eggs carried on skin under fins etc.
      c.   Mouth brooders
      d.   Gill-chamber brooders
      e.   Pouch brooders

2.   Internal bearers

    a.   Facultative internal bearers; occasional internal fertilization of normally oviparous fish, eggs rarely retained long

    b.   Obligate lecithotrophic live bearers; no maternal - embryonic nutrient transfer

    c.   Matrotrophous oophages and adelphophages; one or a few eggs developing at expense of other eggs or embryos.

    d.   Viviparous trophoderms; nutrition partially or entirely from female via "placental" structures

---

While teleosts usually have round eggs, most engraaulids have eggs that are ellipsoidal, thought to be an adaptation to reduce *cannibalism* by the filter-feeding parents after spawning. Other families like the gobies have slightly flattened eggs, and demersal eggs are sometimes irregular in shape. Oviparous elasmobranchs have eggs of unusual shapes (the *"mermaid's purse"*) with tendrils for attachment. Tendrils are also found in the silverside *Atherinopsis* and the gar *Belone*, while the flying fish *Oxyporhamphus* has spines. It is easy to understand the adaptive value of tendrils in distantly related species, but it is much more difficult to explain the ornamentations of the chorion. Most teleosts have a smooth surface to the chorion, but the unrelated inshore dragonet *Callionymus* and bathypelagic gonostomatid *Maurolicus muselleri* have chorions with hexagonal facets, and the flatfish *Pleuronichthys coenosus* has a chorion with very many small facets.

The yolk is usually translucent, unpigmented, and homogeneous in texture, but may be segmented in primitive species like the pilchard *Sardina pilchardus* and sprat *Sprattus sprattus*. In some soleids the segmentation is confined to the periphery of the yolk, and in other species like the jack mackerel *Trachurus symmetricus* segmentation appears progressively during early development. Most commonly, pelagic fish eggs have a single oil globule in the yolk. Of a total of 515 species checked by *Ahlstrom and Moser* (1980), 60% had one oil globlue, 25% had no oil globule, and 15% had multiple oil globules. The oil globule, when single, usually lies at the vegetal pole in Marine fish larvae. It is generally thought that the oil globules are a specialized form of nourishment and have a minimal effect on buoyancy.

As soon as fertilization takes place, the egg absorbs water, the *perivitelline space* forms, and the *chorion* hardens. The perivitelline space is usually narrow but is wide in some *"primitive"* species such as the pilchard, in some eels, and in unrelated species like the stripped bass *Morone labrax* and long rough dab *Hippoglossoides platessoides*. Cleavage is *meroblastic* in hagfish, elasmobranchs, and teleosts, although in the lampreys it is *holoblastic* but with the formation of *micro*-and *macromeres*. In primitive groups like the bowfin *Amia*, gar *Lepisosteus*, and sturgeon *Acipenser*, cleavage is *intermediate* or *semiholoblastic*. The embryo develops as a *blastodisc* at the animal pole. The periphery of the blastodisc overgrows the yolk *(epiboly)*, eventually enclosing it to form a gastrula but leaving an opening, the blastopore. The embryonic axis forms by a process of convergence and concentration in relation to the dorsal lip of the blastropore at the neurula stage but the quantity of yolk influences the timing of such events. The head and eye cups are soon identifiable and the trunk lengthens and separates from the yolk sac. The heart functions well before hatching, and in some demsersal eggs a vitelline circulation can be seen within the yolk sac.

Before hatching, the embryo becomes very active and the chorion is softened as a result of enzymes secreted by *hatching glands*. The degree of differentiation of the newly hatched larva depends very much on the species and egg size, and the incubation period depends on these factors and on temperature. In many marine pelagic species the mouth and jaws are not formed, the eye is not pigmented, the yolk sac is huge, and a primordial finfold runs around the trunk in the median position. Apart from a few melanophores, the larva is very transparent. All newly hatched larvae have free neuromasts on the head and trunk and otoliths are present in the otic capsule. Other marine species hatch with the alimentary system nearly functional and with pigmented eyes. Some larvae are very advanced, and in loricariids the dorsal and caudal fin are partly developed at hatching. In flying fish, flexion of the notochord actually occurs before hatching. In the salmonids — for example, rainbow trout — although the yolk sac is still large, the larva (*alevin*) is better developed and especially the vascular system and vitelline circulation are conspicuous with the blood containing hemoglobin. The young of cichlid and ariid mouth brooders are also further developed and adapted to early life within the parental mouth. In ovoviviparous and viviparous species the young may hatch effectively as postmetamorphic juveniles.

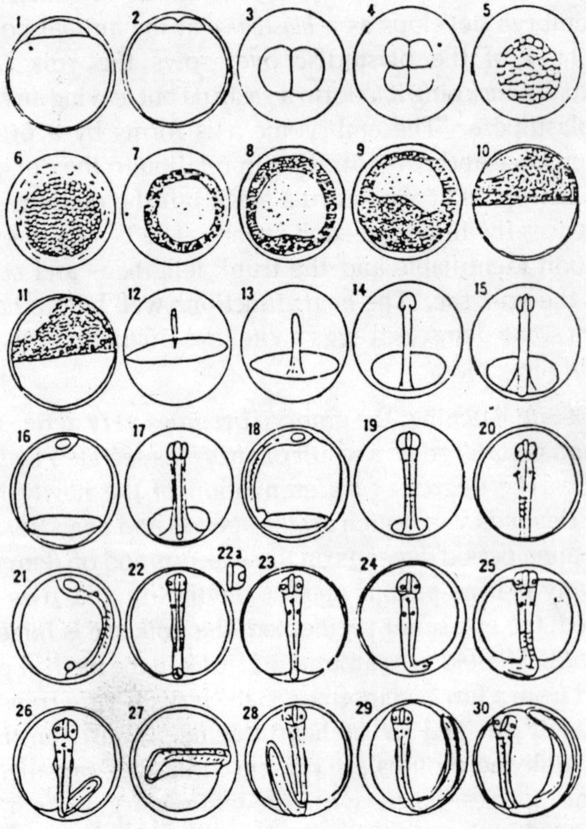

Fig. 2.1.   Development of the dab *Limanda limanda*, using Apstein's stages.

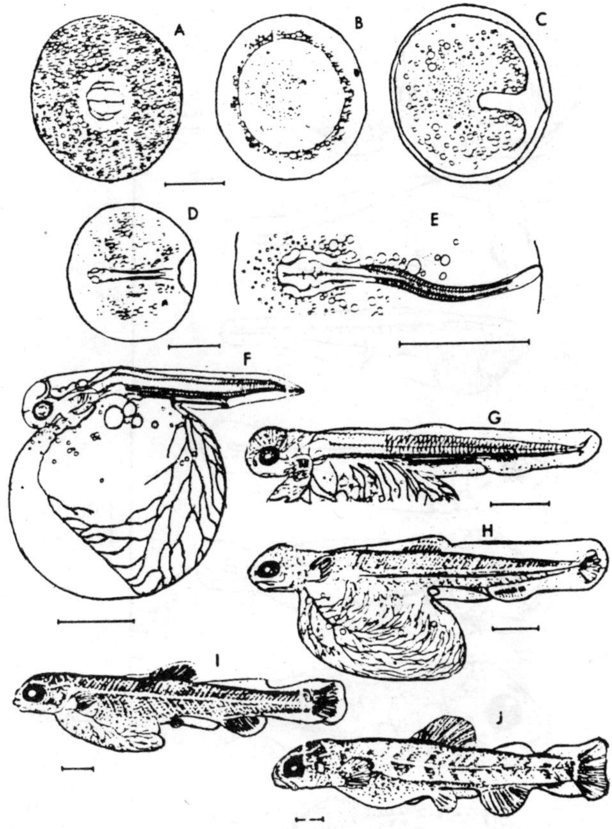

Fig. 2.2. Development of the rainbow trout *Salmo gairdneri*. (A) 8-Blastomeres. (B) Early embryo apparent, one-third epiboly. (C) 0-5 Somites, one-half epiboly. (D) Otic placodes, three-fourths epiboly. (E) Caudal bud with 10-20 somites, total somites 51-58, heart beating. (F) Posterior cardinal veins formed, choroid of eye pigmented. (G) Near hatching, pelvic fins develop. (H) Hatched alevin, first anal and dorsal fin rays. (I, J) Later alevin stages as yolk is resorbed. Scale bars 2 mm long.

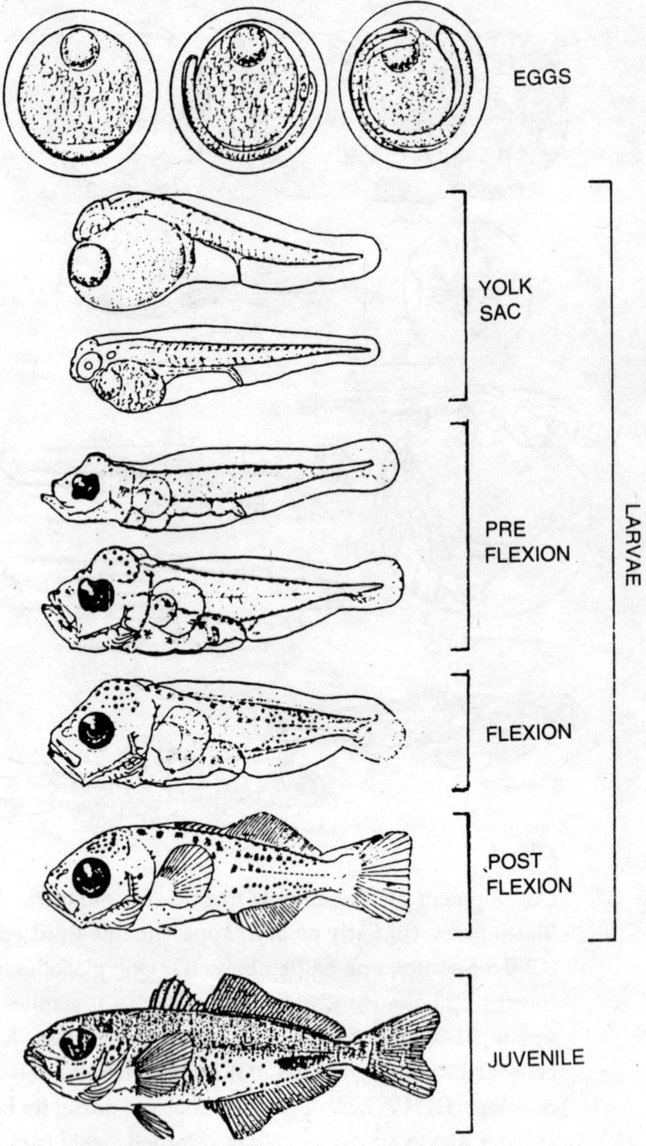

Fig. 2.3. Early life history stages of the jack mackerel *Trachurus symmetricus*.

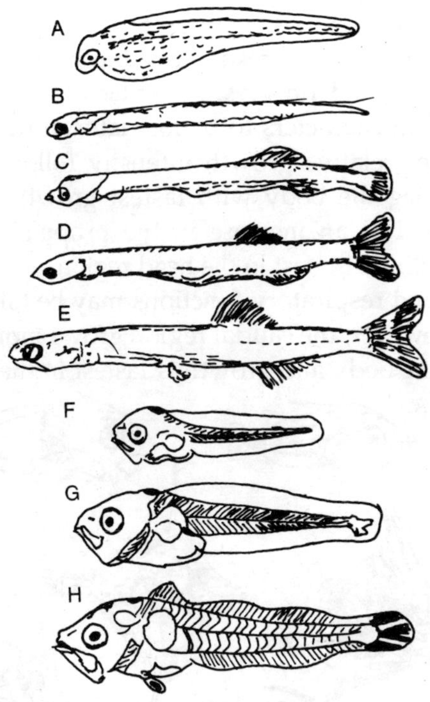

Fig. 2.4.   Development of teleost larvae. (A-E) Northern anchovy *Engraulis mordax*, 2.5, 7.5, 11.5, 18.4, 31.0 mm. (F-H) Pacific hake *Merluccius productus*, 4.3, 7.7, 11.0 mm.

The duration of the yolk-sac period depends on both species and temperature but also on egg size. The argentine *Argentina silus* and the halibut have egg diameters of 3.0 - 3.5 mm. Unexpectedly, the newly hatched larvae are very undeveloped but the halibut takes 50 days to resorb its yolk (at 5.3°C) and reaches a length of 11.5 mm and the argentine reaches a prodigious length of 17 mm on its yolk supply. During the yolk-sac period the mouth and gut and the eyes become functional to allow the larva to switch from endogenous to exogenous nutrition. The subsequent larval period ranges from a few days to some months (and even 2-3 years in eels), depending on temperature and species. During this time the larva is likely at least to double its length and to increase its weight by 10 to 100 times. Transient characters, such as spines, may appear which are presumably antipredator adaptations. Eyetalks, elongated fin rays, or tentacles, may also appear, often as larval characters to be lost later in development. In some species relative growth intensity follows a U-shaped gradient along the body with fastest growth in the caudal region, linked to an increase in the propulsive area of the body. Growth is also fast in the head region, where elaboration of feeding and respiratory functions may be taking place. In sculpins, however, the caudal region grows more slowly than the rest of the body and growth is fastest in the head region.

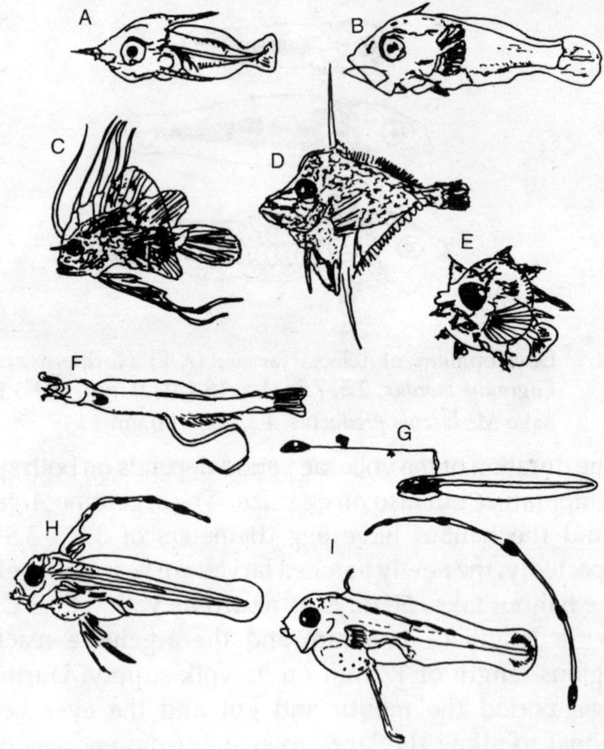

Fig. 2.5. Teleost larvae showing spines and other processes. (A) *Holocentrus Cexillarius* 5.0 mm. (B) Sebastes macdonaldi 9.0 mm. (C) *Lophius piscatorius* 26 mm. (D) Acanthurid 7 mm. (E) *Ranzania laevis* 2.8 mm. (F) *Myctophum aurolaternatum* 26 mm. (G) *Carapaar acus 3.8 mm. (H) Trachipterus sp. 7.6 mm. (F) Zu cristatus 6.5 mm.*

Progressive differentiation of adult characters such as fin rays and skeleton occurs. The larvae eventually pass through a process of *metamorphosis* to the juvenile stage. This process may be rather abrupt or it may be prolonged. Typically the blood becomes pigmented, scales and pigment appear on the body surface, the *meristic characters* such as fin rays are complete, and the body shape becomes like the adult. The juvenile appears as a small adult. In flatfish, *metamorphosis* is a remarkable process as the fish starts to change from the bilaterally symmetrical larva to an asymmetrical juvenile lying on one (*abocular* or *blind*) side. Changes take place to the skull and sense organs and, in particular, the eye of the abocular side migrates across the top of the skull. The young of many elasmobranchs, with a long incubation period, effectively hatch as juveniles, albeit with a yolk sac. In species with parental care the early larvae may also be advanced or "precocial." This variety makes it difficult to categorize early life histories in a neat and convincing way.

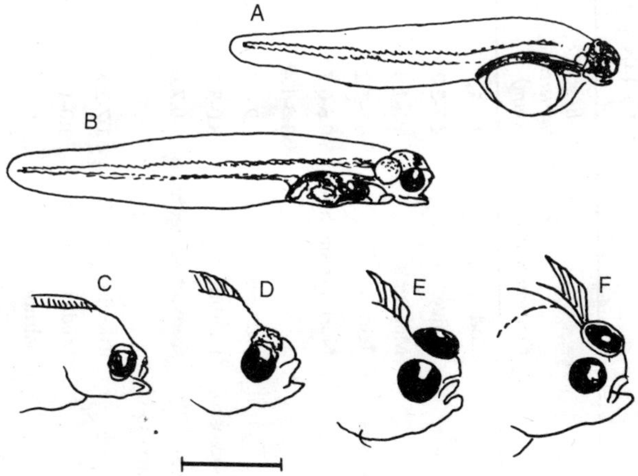

Fig. 2.6. Development and metamorphosis in the plaice *Pleuronectes platessa.* (A) Yolk sac larva, 6.6 mm. (B) Larva at first feeding, 7.3 mm. (C-F) Stages of metamorphosis with eye migration; scale bar 2 mm.

TABLE II

Early Life History Characteristics[a]

Species	Common name	Egg diameter (mm)	Hatching length (mm)	Weeks from fertilization to			Temperature range (oC)
				Hatch	First feed	Metamorphosis	
Gadus morhua	Cod	1.1-1.9	3-4	1.4-2.9	1.9-3.6	4.2-5.0	4-12
Pleuronectes platessa	Plaice	1.7-2.2	6-7	2.5	4.0	10-12	7-11
Scomber scombrus	Mackerel	1.0-1.4	3-4	0.8-1.5	1.3-2.0	11-13	9-15
Scophthalmus maximus	Turbot	0.9-1.2	2-3	0.6-0.9	1.1-1.7	5-6	13-18
Engraulis mordax	Northern anchovy	0.6-0.7 x1.3-1.4[b]	2-3	0.4-0.5	0.9-1.1	11-15	13-15
Clupea harengus	Herring	0.9-1.7	5-8	1.0-3.0	2.0-4.5	12-24	6-14
Hippoglossus hippoglossus	Halibut	3.0-3.2	6-7	2.0-3.0	9.0-10.0	?15-16	4-7
Acanthurus triostegus	Convict surgeonfish	0.7	1.7	0.15	0.7	?	26
Oreochromts (=Tilapia) mossambicus	Tilapia	1.7-2.2	4	0.6-0.7	1.7-2.6	?	28
Oryzias latipes	Medaka	1.0-1.3	4-5	1.5-2.0	2.0-2.4	Not elearent	20-25
Salma salar	Salmon	5-6	15-25	20-22	26-28	Not elearent	1-7
Scylitorhinus caniculus	Spotted dogfish	65 (long)	100	24-32	28-36	28-36	4-12
Squalus acanthias	Spur dogfish	24-32	240-310	-104	-104	<104	4-12

TERMINOLOGY OF EARLY LIFE HISTORY
STAGES

A good terminology should be as simple as possible and linked to both form and function. The production of a generally accepted terminology is a current issue in ichthyology. The difficulties lie in producing a terminology that embraces all species and all patterns of development in fish, and it almost becomes an intellectual challenge to achieve this. Some workers favour a large number of stages, others very few; one point of view suggests terminology based on size alone, another that ecological considerations should be paramount. Some workers use the term *"embryo"* to cover the period from fertilization to first feeding and consider *hatching* to be a relatively insignificant process. While it is certainly true that the change from endogenous to exogenous food supply is a major hurdle for the organism to overcome, it should not be forgotten that eggs cannot avoid predators although hatched larvae can. Many species of fishes hatch in a very well developed state, especially where *ovoviviparity, viviparity,* or other parental care is involved, or where the incubation period within the egg is long; other species hatch in a much earlier state of development. It is difficult to resolve a nomenclature to cover such a wide variation in ontogeny.

In the present we prefer to use the term *"embryo"* only to the point of hatching, does not accept the terms *"prelarva"* and *"post-larva,"* which suggest stages before and after a larval stage, and uses the term *"larva"* to cover development from hatching to metamorphosis and the term *"juvenile"* from metamorphosis to first spawning. Terms such as *"fingerling"* or *"young-of-the-year"* are unsatisfactory: the former can hardly be applied to very short fish or the latter to species with a short generation time. A simplistic approach to terminology may well require additional qualification to be given, such as *"yolk-sac"* larva, or it may have to be made clear that some species hatch in an advanced state of development. This is in broad

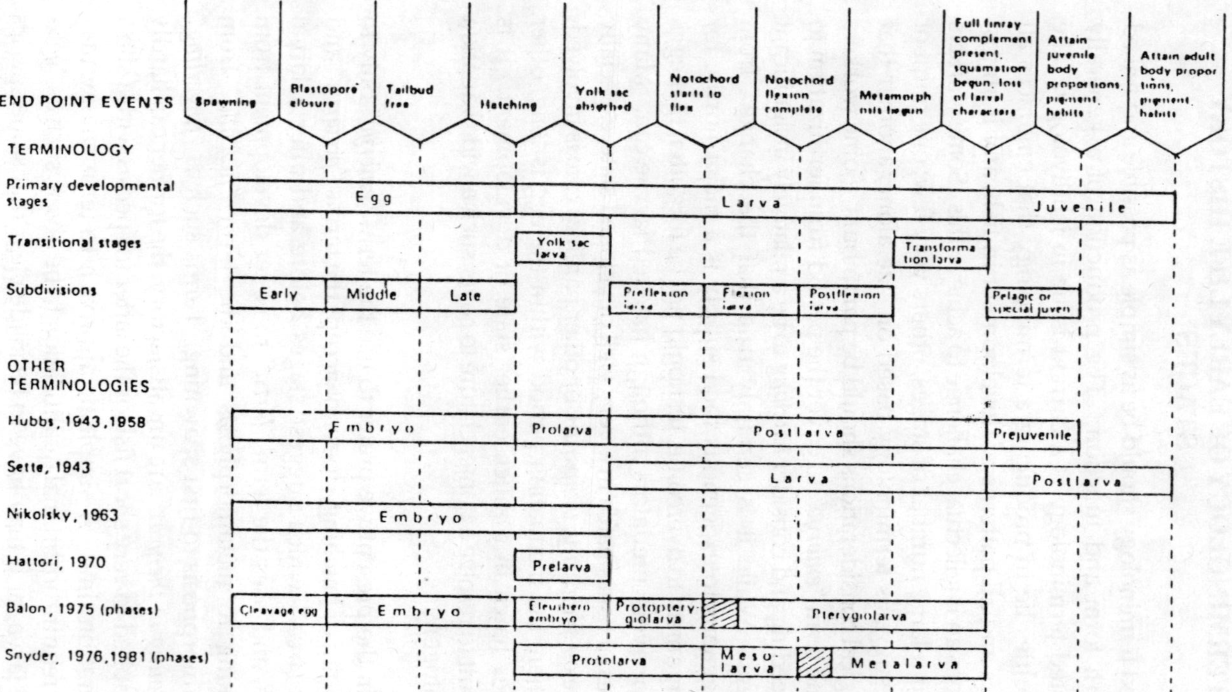

Fig. 2.7.   Terminology of life history stages.

agreement with *Kendall et al.* (1984), who also favour dividing the larval stage into *"preflexion," "flexion"*, and *"postflexion"* substages, referring to the turning up of the notochord tip during the first stages of development of the caudal fin.

EGG SIZE AND EGG QUALITY

Egg Size

The size of eggs is variable having important ecological implications. Although a large yolk sac may reduce locomotor performance, plentiful yolk is likely to extend the period for switching from endogenous to exogenous feeding. Any factor that might provide larger larvae is likely to reduce the dangers from predation, since large larvae are likely to be able to escape faster and to be less susceptible generally to small predators. Furthermore, any factors that might cause early improvements in growth and size are likely to reduce the range of predators more rapidly.

Egg size also influences the rate of development, those species with large eggs having a longer incubation time. While the small eggs of many fecund marine species hatch within a few hours or days, salmonids may take several weeks to hatch, and the very large eggs of elasmobranchs, several months. The ovoviviparous dogfish *Squalus acanthias* has a gestation period of about 2 years.

Egg size does not seem to affect incubation time (at any one temperature) either in Atlantic herring, Atlantic salmon, Arctic charr *Salvelinus alpinus*, tilapia Oreochromis (=*Tilapia*) *mossambicus*, or the orangethroat darter *Etheostoma spectabile*. However, egg size does influence larval size, with larger larvae, at first feeding, being produced from larger eggs. Furthermore, the larvae from larger eggs live longer on their yolk reserves although the effect on ultimate survival may not always be established.

Wallace and *Aasjord* (1984) found the effect of egg size on the length of Arctic charr alevins was still clearly evident 140 days post-hatching; *Springate et al.* (1985) also found that the greater size of rainbow trout fry from females fed on a high ration could be followed for 2-3 months *Glebe et al.* (1979) showed that the length of fry of Atlantic salmon from four New Brunswick rivers was still related to the original egg diameter some 8 months later. *K.J. Rana* reared tilapia and found that the weight of 20-, 40- and 60-day-old fish was significantly related to the original egg size, although the degree of association decreased over the 20- to 60-day period. *Thorpe et al.* (1984) and *Springate* and *Bromage* (1985), concluded that the benefits of large egg size were soon lost during subsequently growth.

Intraspecifically, there are also many factors that control egg size. These are:

Parental Size

Larger females produce larger eggs in Atlantic salmon *Salmo salar*. In addition, females that spend longer at sea (and so are larger) also produce larger eggs. In Atlantic herring, *Hempel* and *Blaxter* (1967) found that the eggs of first-time spawners were the smallest but that there was no influence of parental size on egg weight in repeat spawners.

Spawning Group

There is an extensive literature showing how salmonid eggs vary in size from different rivers. For example, *Glebe et al.* (1979) measured the calorific value of Atlantic salmon eggs from four New Brunswick rivers and found a range from 1067-1576 J/egg.

Season

The influence of spawning group is closely linked to different spawning seasons that may occur within a species. This is most dramatically shown in the herring, where different seasons may have egg dry weights varying by a factor of four, with the largest eggs being produced in the winter and spring and smallest in the summer and fall. This general intraspecific trend toward smaller eggs as the spring season progresses was shown in a wide range of species.

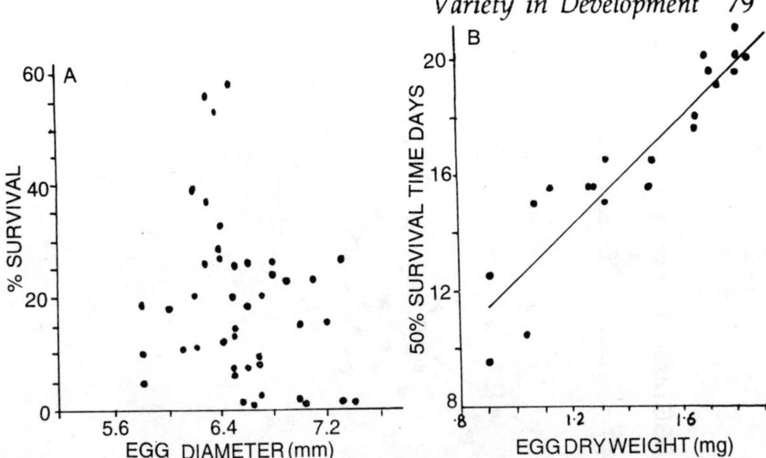

Fig. 2.8.   (A) Percent survival of Atlantic salmon parr *Salmo salar* from eggs of different diameter. Each point refers to a different female. There is no significant correlation between egg size and survival. (B) The time to 50% survival from hatching of unfied tilapia *Oreochromis (Tilapia) mossambicus* fry related to the mean dry weight of the eggs. Each point refers to a different clutch of eggs. The regression is significant ($r= 0.923$, d.f. = 23, $p < 0.01$).

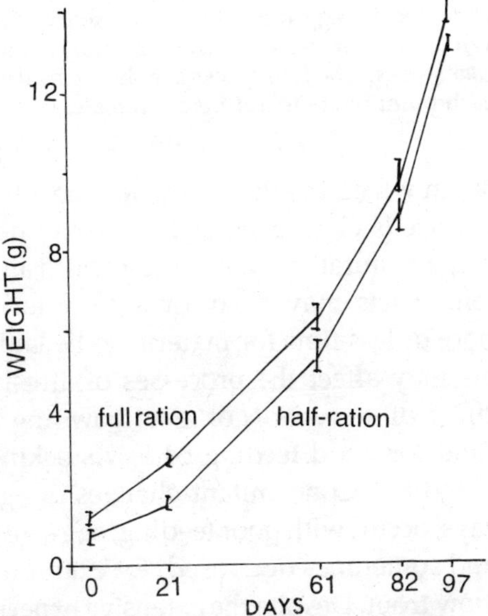

Fig. 2.9.   The weight (mean =S.E.) of rainbow trout *Salmo gairdneri* fry derived from eggs of females held on half and full rations. Day 0 refers to first feeding, so that the difference in weight is still detectable 97 days later.

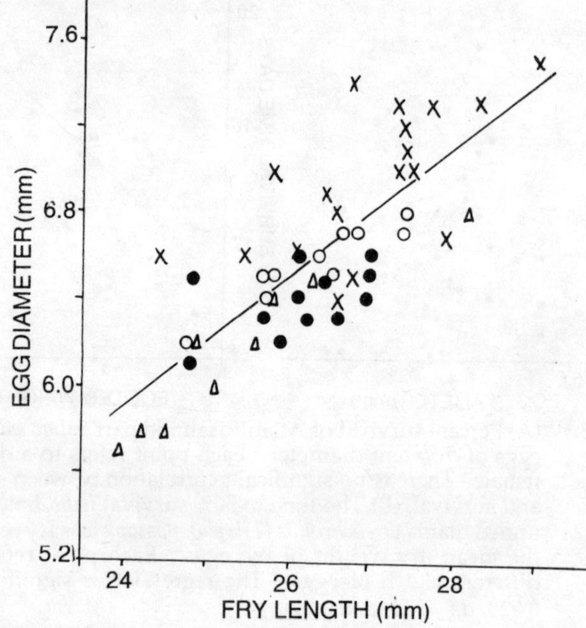

Fig. 2.10. The effect of egg diameter on subsequent length of fry in four New Brumswick (Eastern-Canada) Atlantic salmon *Salmo salar* stocks. The four stocks are shown by different symbols. Each point refers to a different female.

Diet

The way in which the diet of the female affects fecundity and egg size needs clarification. Much may depend on the phase of the egg maturation cycle during which an experimental diet is given. Diets may delay or accelerate spawning, so allowing more or less time for material to be laid down in the egg, or diets may affect the processes of atresia. The most common effect of starvation or overcrowding is to reduce fecundity, and for good feeding or low stocking density, to increase fecundity. Concomitant changes of egg size often, but not always, occur; with poor feeding, egg size is sometimes increased and sometimes decreased, for example, in haddock and in rainbow trout. Despite the extensive experimental work on egg size, it is not clear the extent to which changes of fecundity or egg size have adaptive value in the wild and whether egg survival can be enhanced or not.

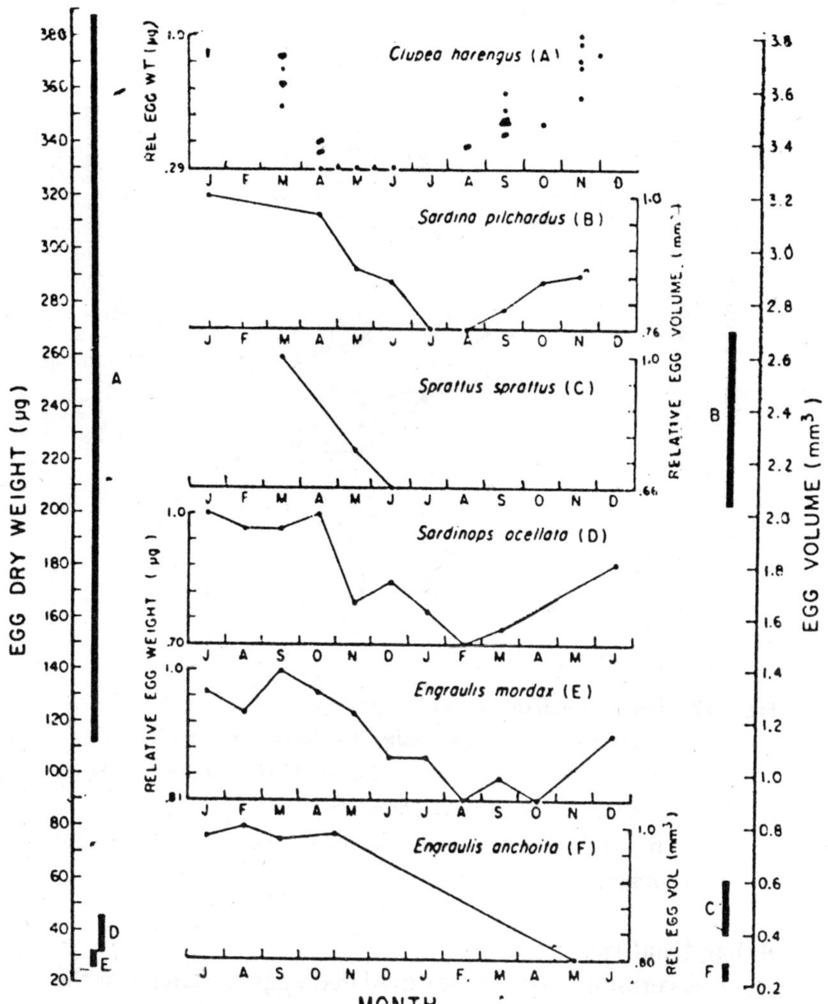

Fig. 2.11. Relative changes of egg weight or volume in different spawning seasons (center graphs) where 1 = maximum egg weight or volume for the season. Species from the southern hemisphere (*S. ocellata* and *E. anchoita*) are offset by 6 months. The separate ordinates show the range in absolute weight or volume of eggs. Note the enormous range of egg weight in herring spawning in different seasons compared with other species.

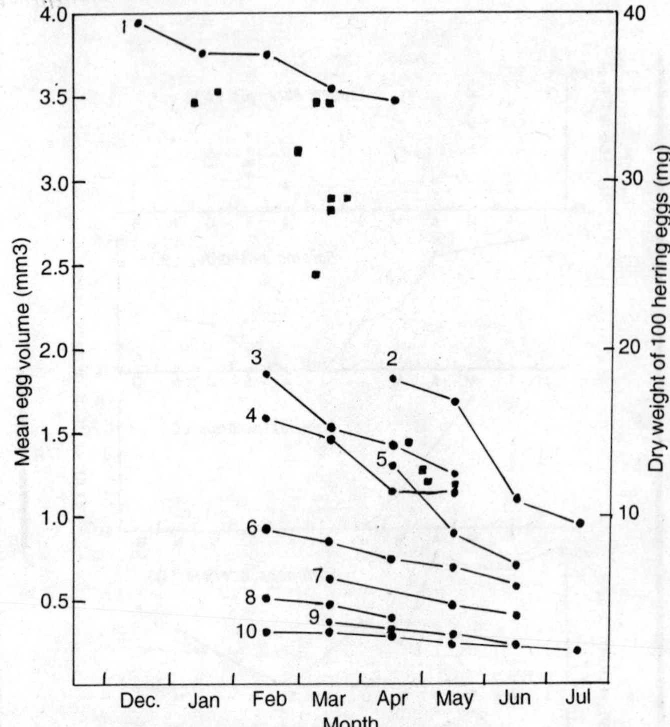

Fig. 2.12.   The seasonal decrease in mean egg volume of (1) *Pleuronectes platessa*, (2) *Trigla gurnardus*, (3) *Melanogrammus aeglefinus*, (4) *Gadus morhua*, (5) *Solea solea*, (6) *Merlangius merlangus*, (7) *Sprattus sprattus*, (8) *Platichthys flesus*, (9) *Rhinonemus cimbrius*, and (10) *Limanda limanda*. Square symbols show dry weight of herring eggs.

B. Egg Quality

It is certainly possible for ovulated eggs to remain in the body cavity of the female for many days, often culminating in their resorption into the maternal tissues. *Craik* and *Harvey* (1984b, c) found that the eggs of rainbow trout *S. gairdneri* held for 30 or more days postovulation underwent changes of water content and biochemical components that must have reflected a deterioration and decomposition of the yolk. They also found that the percentage of eggs that hatched fell sharply when the eggs were held for 18 days postovulation. *Springate et al.* (1984) found maximum egg and fry survival if rainbow trout were stripped 4-6 days after ovulation. In contrast, *Mollah* and *Tan* (1983) found that the eggs of catfish *Clarias macrocephalus* only remained viable for up to 10 h postovulation at 26-31°C. The need to fertilize eggs at an optimal time is particularly important in cases of hand stripping, or if spawning is induced by hormones. Flounder *(Limanda yokahamae)* eggs remain in good condition for 2-3 days after ovulation, depending on the hormone used for inducing spawning.

The decline in egg viability at postovulatory stage seems much more significant than small differences in dry weight, protein, and lipid content of eggs from different rainbow trout females, which may also be related to hatching success. Possibly a more important biochemical component of salmonid eggs is the carotenoid. They may act as precursors of vitamin A and play a part in respiration; they are certainly a source of pigment for the chromatophores. It is likely that they have to be above a critical concentration to give high hatch rates. Variation in egg quality may arise in other ways. Adverse feeding, temperature, light, and water quality, as well as crowding can cause low fertilization and hatching rates. In the red sea bream *Chryosophrys major* a low-protein, phosphorous-deficient diet produced eggs with poor hatchability and deformed hatched larvae. In other species, such as the rainbow trout, ayu *Plecoglossus altivelis,* and carp *Cyprinus carpio,* deficiencies of fatty acids and vitamin E in the diet of the female may effect egg viability. *Whipple et al.* (1981) found that striped bass migrating through polluted water showed poor condition of the parents, low fecundity, and low egg viability, as well as a high level of lesions and parasites. Poor-quality eggs may show irregular shape, abnormal fertilization, soft chorions, or negative buoyancy. In a unique study of a wild stock, *Kjorsvik et al.* (1984) found that 6-60% of cod eggs had abnormal mitoses or chromosome aberrations.

THE EFFECT OF STARVATION

A number of morphological and chemical changes occur during starvation. There is a progressive collapse of the larval head and trunk in Atlantic herring and plaice and in jack mackerel, so that body weight or head height relative to the length can be used as an index of starvation. The condition factor (dry weight/length^3) has been used on sea-caught Atlantic herring larvae to assess their nutritional condition. A U-shaped relationship is found between condition factor and length because the larval body is denser when yolk is present and later as ossification occurs. There were also changes in biochemical components in plaice and herring larvae during starvation. Percentage of water increased and percentage of triglyceride, carbohydrate, and carbon decreased. Percentage nitrogen decreased in plaice but not in herring. Recently *Buckley* (1981) and *Clemmesen* (1987) have shown that the RNA/DNA ratio drops dramatically in starving winter flounders, herring, and turbot larvae. The time over which starvation takes place depends on a number of factors. *Theilacker* and *Dorsey* (1980) and *McGurk* (1984) summarized the time to reach the point-of-no-return, or irreversible starvation, in some 25 marine species if they failed to feed once the endogenous yolk supply had been exhausted. There were very large differences, ranging from 3.5 days postertilization in *Anchoa mitchilli* at 28°C to 36.5 days postertilization for Pacific herring at 6°C. Generally the time to the PNR for first-feeding larvae increases for species with long incubation periods and at low temperatures and when the eggs are large. Once feeding is established the time to the PNR increases with age, as shown in Fig. 2.14, presumably because of greater body reserves.

Although the morphological and chemical effects of starvation during development are now fairly well established, it is not certain how much these changes are reflected in locomotor performance and behaviour. *Blaxter* and *Ehrlich* found a decrease in sinking rate (or an increase in buoyancy)

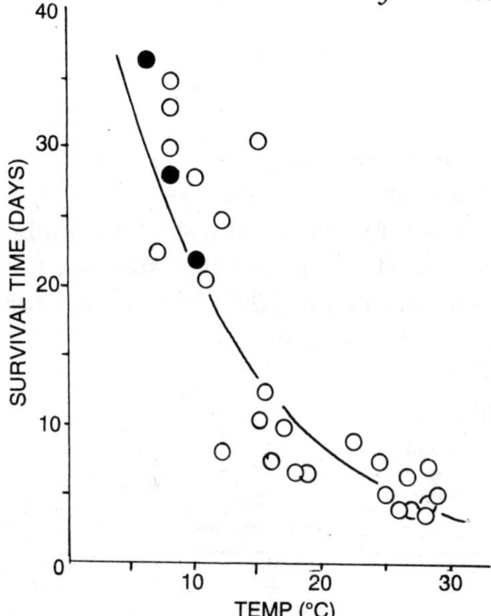

Fig. 2.13. The time from fertilization to irreversible starvation (PNR) of unfed fish larvae of 25 species at different temperatures (open circles). The black circles are larvae of Pacific *Clupea pallasi*.

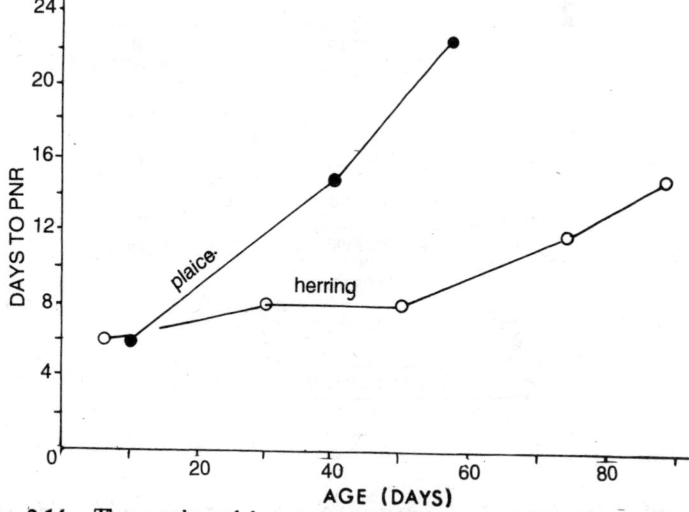

Fig. 2.14. The number of days to irreversible starvation (PNR) of different aged plaice *Pleuronectes platessa* and Atlantic *herring Clupea harengus* larvae. Temperatures 7.5-11.5°C.

in plaice and Atlantic berring larvae as a result of increasing hypotonicity of body fluids and a decrease in body protein during starvation. Activity was maintained for some days and tended to decrease drastically only late in the period of starvation, presumably as a way to maintain food-searching ability as long as possible, even if it was energetically expensive. *Huse* and *Skiftesvik* (1985) found that starving turbot larvae swam faster for shorter periods and searched for food less efficiently than feeding larvae.

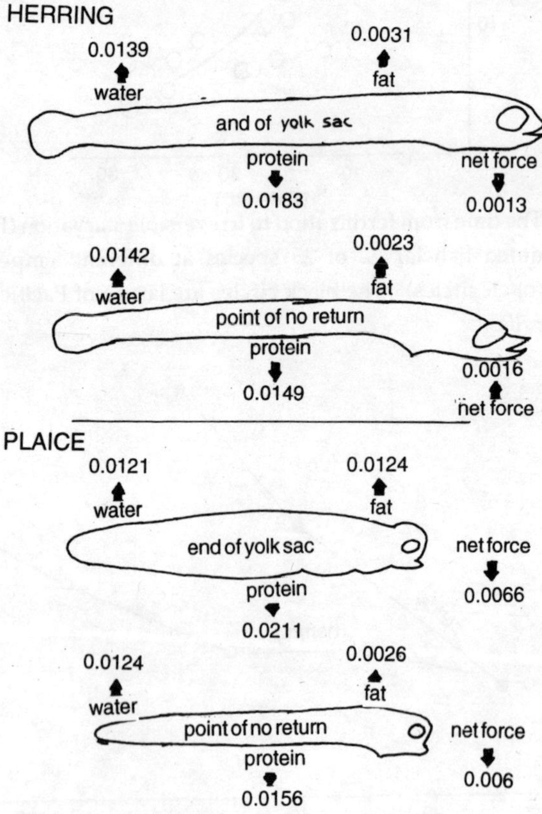

Fig. 2.15.   Buoyancy forces due to chemical components in Atlantic herring *Clupea harengus* and plaice *Pleuronectes platessa* larvae at the end of the yolk-sac stage and at the point-of-no return

(PNR). All forces are dynes/wet weight (mg); arrows show direction of forces. Note that the larvae become increasingly buoyant as they starvae.

THE EFFECT OF CAPTIVITY

Whether size hierarchies (sometimes called depensation of growth) occur in natural conditions is uncertain, since variations in egg size and spawning time may obscure the phenomenon, and selective predation on small individuals may limit the lower end of the size range. The increase in length range with age may be due to a natural variation early in development.

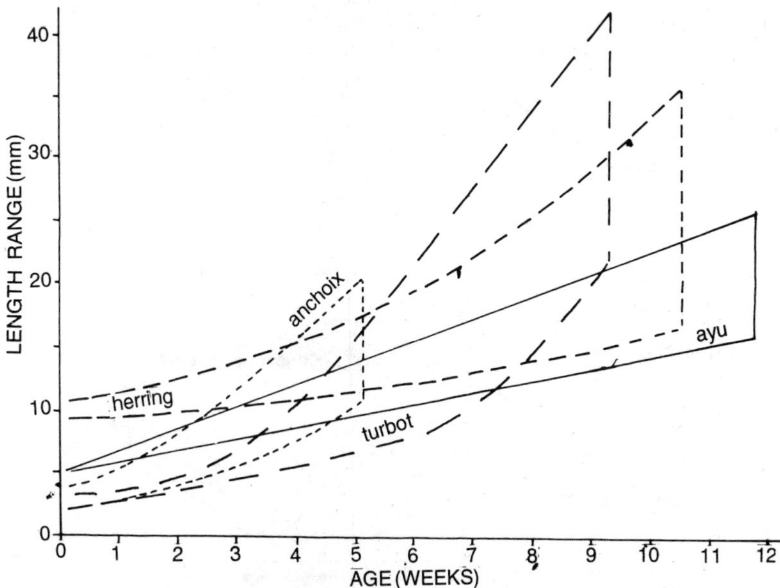

Fig. 2.16. Size hierarchies. Increase of length range with age in northern anchovy *Engraulis mordax*. Atlantic herring *Clupea harengus* turbot *Scophthalmus maximus*, and ayu *Plecoglossus altivelis* reared in the laboratory.

For example, a 10% variability in length of 10-11 mm early in development would result in a range from 50 to 55 mm later. The increase of range in length is often well beyond such considerations, and it seems likely that size hierarchies are a tank phenomenon caused by competition for food, social dominance of some individuals in crowded conditions, and lack of predation on smaller individuals. Thus *Eaton and Farley* (1974) found a reduction in the size hierarchy of the zebrafish *Danio (=Brachydanio) rerio* fed an adequate diet. After 32 days, larvae fed on a low ration ranged from 5 to 12.5 mm (mean 9.5 mm), and after 39 days another group fed on a high ration ranged from 16.0 to 17.5 mm (mean 16.9 mm).

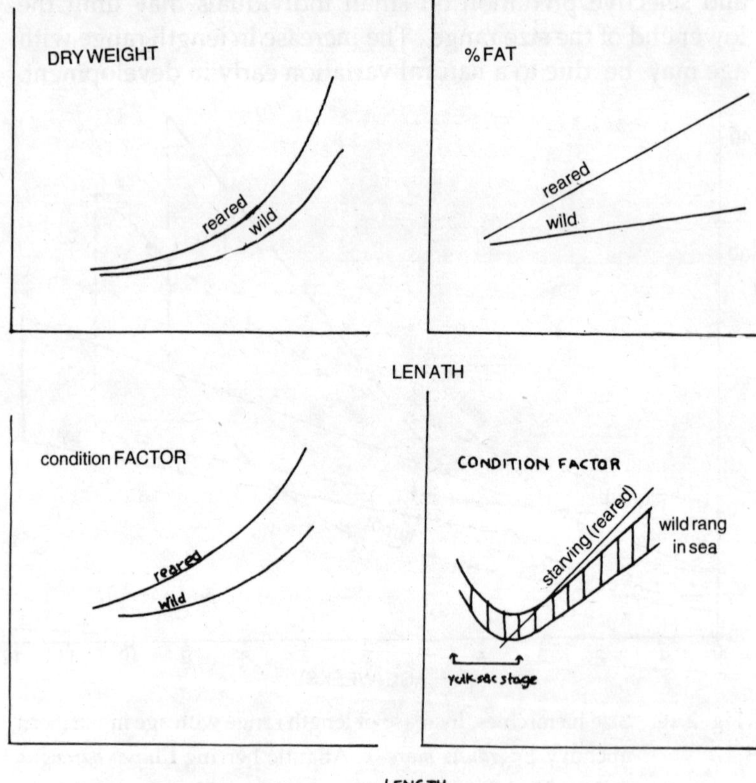

Fig. 2.17.   A general comparison of changes in dry weight, fat, and condition factor of reared and "wild" fish as they grow, based mainly on Atlantic herring; see text for details. The change of condition factor (shown bottom right) in the sea results from initial loss of yolk, causing a fall, followed by

> progressive ossification, causing a rise. Condition factors of reared starved larvae tend to be higher than "wild" larvae, showing that starvation criteria cannot be obtained satisfactorily from captive larvae.

Rearing in captivity also tends to produce shorter, fatter fish, with high condition factors and growth abnormalities such as foreshortened snouts, neoplasms of the head, and failure of eye migration and fin development in flatfish. *Arthur* (1980) found that the hearts of northern anchovy larvae, as determined by length of the ventricle, were as much as 40% longer in laboratory-reared compared with sea-caught individuals. Excessive contact with the walls of the tank, other forms of stress, and social factors can be implicated in causing these abnormalities.

The effect of the volume of the rearing container was also investigated by using jack mackerel larvae reared at the same density in 10-1 and 100-1 circular tanks. The beneficial effect of the larger volume was apparent after only 4-5 days of feeding, when the larvae were larger and in better nutritional condition. The dramatic effect of using rearing facilities on the mesocosm scale was shown by *Morita* (1985) and *Paulsen et al.* (1985), who reported excellent growth and survival of Pacific herring larvae and turbot larvae respectively in 20-m^3 on-shore tanks. *Sturmer et al.* (1985) found similar beneficial effects using red drum larvae in large on-shore tanks using cod larvae in a coastal impoundment of 60,000 m^3. High density of larvae may also create problems other than those associated with physical interactions. Crowded conditions may cause the production of inhibitory substances. Although originally demonstrated in amphibians, such substances may be implicated in population control of the guppy *Lebistes (Poecilia) reticulatus*, whitecloud maintain minnow *Tanichthys albonubes*, tiger barb *Barbus tetrazona*, rainbow trout, and ciscoes *Leucichthys sp. Laale and McCallion* (1968) found that the development of zebrafish could be arrested before gastrulation by the use of supernatant homogenates produced from other zebrafish embryos. Such effects are likely to be relevant in the wild only to species that live in crowded conditions or in stagnant water.

In biochemical terms, hatchery fishes often have a higher percentage fat content than their wild counterparts but lower percentage protein and ash. The biochemical and morphological changes occurring in hatchery fishes can be reflected in their behaviour and especially in their locomotor performance. Crowding induces stress and aggression that can be partly alleviated by adequate food, for example, in the medaka *Oryzias latipes* and Atlantic salmon. *Bams* (1967) found that the fry of wild migrant sockeye salmon *Oncorhynchus nerka* could stem a water current and avoid predation better than hatchery fry, although mainly this was due to the fact that they were bigger.

Developing fishes may not be subjected to the normal interplay of light and shade, nor to the very high light intensities appertaining in the wild. They may not be able to practice avoidance or other responses in the absence of a natural substratum, typical water currents, or predators.   Such deprivation will be particularly harmful where learning is involved in the development of behaviour patterns.

A number of workers have assessed the effect of blinding, or rearing in darkness, on the eye and optic tectum of fish larvae. *Pflugfelder* (1952) found that unilateral blinding of newly hatched swordtails *Xiphophorus* and guppy *Lebistes* caused a reduction in the development of the contralateral optic tectum, mainly by a decrease in volume of the ganglion cells. Experiments in darkness confirmed that the effect was caused by lack of visual input rather than by degeneration products released from dying axons of the optic tract. *Blaxter* (1970) could find no retinal degeneration in dark-reared Atlantic herring larvae. *Zeutzius et al.* (1984) found that dark-rearing of tilapia *Sarotherodon (=Tilapia) mossambicus* did not affect the normal outgrowth of the nerve fibers of the optic tectum into the retina. It did, however, reduce the optic layer and the differentiation of the synapses, where the number of synaptic vesciles increased.

Dark-rearing can affect behaviour. *Blaxter* (1970) found that newly hatched Atlantic herring larvae were very inactive when returned to the light after rearing in the dark and subsequently showed a high mortality. *Zeutzius* and *Rahmann* (1984) found that dark-reared larvae of tilapia failed to swim up after yolk resorption; visual acuity was impaired after 20-30 days in the dark, and after 50 days no optokinetic nystagmus was present. Effects no body weight and length were minimal, although the increase in body depth was substantial compared with control fishes.

RATE OF DEVELOPMENT

The rate of development is clearly under genetical control. Within a species it is most strongly influenced by temperature, and this is exemplified by data on days to hatch at different temperatures in 13 species. *Blaxter* (1969) and *Herzig* and *Winkler* (1986) discuss the mathematical relationship between temperature and incubation and the use of Q_{10} and other temperature coefficients. In particular Q_{10} is found to vary with range of temperature, and it may be that optimum temperatures for development, where hatching rates are highest, take place where the Q_{10} is between 2 and 3.

Temperature can also influence size at hatching, effeciency of yolk utilization, growth, feeding rate, time to metamorphose, behaviour and swimming speed, digestion and gut evacuation rates, and metabolic demand. Temperatures also has indirect effects on larvae through the oxygen capacity of water, is viscosity, and phytoplankton blooms.

The effect of temperature on the interaction of differentiation and growth has been studied. Salmonids and clupeids incubated at high temperatures tend to weigh less at hatching. The best known effect is on meristic characters — counts of serial

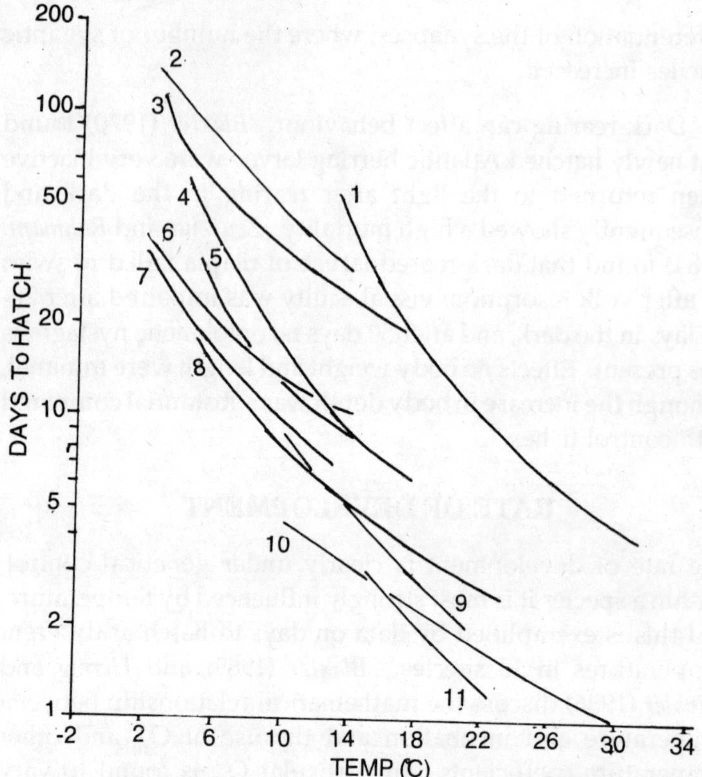

Fig. 2.18. The time from fertilization to hatching for 11 species of teleost related to temperatures. Data originally cited in Blaxter (1969) except where stated. (1) Desert pupfish *Cyprinodon macularius*, (2) brook trout *Salvelinus fontinalis*, (3) rainbow trout *Salmo gairdneri*, (4) smelt *Osmerus eperlanus*, (5) Atlantic herring *Clupea harengus*, (6) plaice *Pleuronectes platessa*, (7) Pacific cod *Gadus macrocephalus*, (8) rockling *Enchelyopus cimbrius*, (9) mackerel Scomber scombrus, (10) grey mullet *Mugil cephalus* and (11) striped bass *Morone saxatilis*.

structures such as vertebrae, fin rays, scales, and gill rakers, which are labile (within limits) and susceptible to various environmental factors. Until recently it has not been clear what adaptive advantage might exist for varying numbers of meristic characters. It has been shown that individuals of a particular species with more vertebrae are often longer, and it seems plausible that they might also be more flexibile and able to swim faster. *Swain* and *Lindsey* (1984) have recently shown that there was selective predation for vertebral number in young sticklebacks *Gasterosteus aculeatus* preyaed on by sunfish *Lepomis gibbosus*. The survival of 8.2-mm-long sticklebacks was 1.3-1.7 times greater for fish with 31 vertebrae than with 32. This effect was not found with 8.9-mm-long sticklebacks, nor was there any influence of temperature.

Early development can be enhanced hormonally. *Dales* and *Hoar* (1954) treated the eggs of chum salmon *O. keta* with thyroxine and thiourea, an antithyroid compound. Thyroxine accelerated growth of the body wall and pectoral fins, increased guanine deposition, decreased pigmentation, caused exophthalmia, but reduced the rate of increase of body length. Thiourea decreased guanine deposition and also decreased the rate of growth in length. More recently, it has been shown that immersion in thyroxine accelerates growth in larval tilapia *Sarotherodon* (= *Tilapia*) *mossambicus*, carp *Cyprinus carpio*, and milkfish *Chanos chanos* and enhances survival in tilapia and carp.

ORGAN SYSTEMS

A. Alimentary System

Many species hatch without a mouth, but this develops rapidly to allow for the transfer from endogenous yolk to exogenous food. Feeding in many species is a predatory act requiring vision, and feeding does not occur in the dark, especially in the very young stages. The size or gape of the mouth at first feeding, and therefore the size of food that can be taken, is crucial for survival at the end of the yolk-sac stage. Experiments show that the size of food taken is related to the gape of the jaw and that both increase with age of the larvae.

Length and complexity of the alimentary tract increase as

the larvae grow. *Dabrowski* (1984) describes three groups of fish larvae based on the morphology of the alimentary tract and gut enzymes. Most species have early larvae without a functional stomach or gastric glands; salmonids, on the other hand, have a functional stomach before changing to external feeding. *Tanaka* (1973) gives a full account of the development of the alimentary tract in 21 Japanese marine and freshwater species. At first feeding there are no pharyngeal teeth and few, if any, taste buds. The esophagus has longitudinal folds and mucous cells; the intestine and rectum are lined with columnar epithelium, and it is likely that most digestion occurs here. Cilia are present in the gut of early clupeoid and salangid larvae and may help to pass food along the gut. The liver, gallbladder, and pancreas are also formed early. It many species, but not in salmonids like the rainbow trout, the stomach and pyloric caeca develop late in larval life as the pattern and quantity of feeding changes. These subsequent processes are well described in the northern anchovy, and in the spot *Leiostomus xanthurus*.

The length of the gut influences the passage time for food. For example, in roach *Rutilus rutilus* larvae the food is retained for only 2.5 h at 20°C, whereas in the adult it is 6 h. The time for digestion and resorption and for the recovery of digestive enzymes is therefore reduced in the larval stage. *Dabrowski* (1984) reviewed work on the appearance of digestive nzeymes during development. *Clark et al.* (1985) found an increase in protease activity of Dover sole from the age of 24 days up to the adult stage. Similarly a progressive increase in activity and number of proteolytic enzymes have been found with age in whitefish *Coregonus* hybrids, in rainbow trout, and in roach. Higher proteolytic activity in the roach could be correlated with the lack of a stomach. In the rainbow trout the well-developed digestive tract, which is differentiated into a stomach, pyloric caeca, and a short intestine at first feeding, may compensate for a lower level of proteolytic activity.

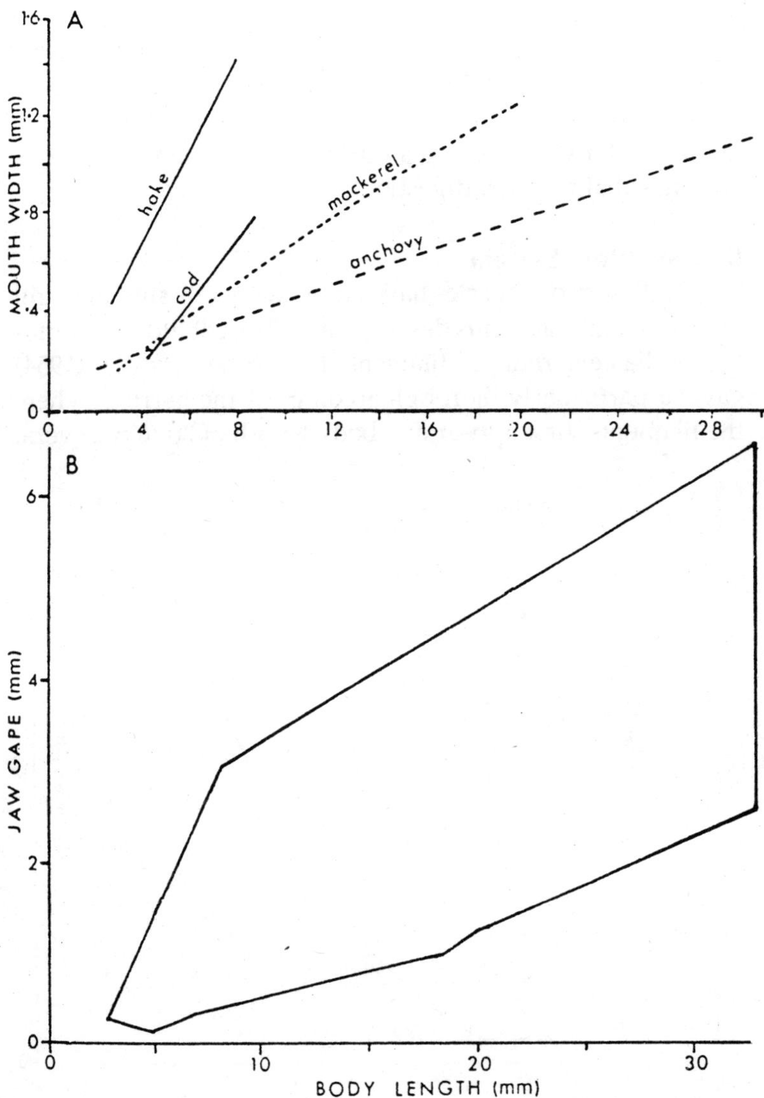

Fig. 2.19. (A) The relationship between body length and width of the mouth in the larvae of hake *Merluccisu merluccisu, cod Gadus morhua*, Pacific mackerel *Scomber japonicus*, and anchovy *engraulis ringens.* (B) "Profile" of the body length: vertical jaw gape relationship in the larvae of 19 teleost species.

The relatively underdeveloped state of the alimentary system may explain why most species have carnivorous larvae, although later, when the gut lengthens, they may become herbivorous. It may also explain why artificial food is less satifactory for young larvae, since live food contains exogenous enzymes that may aid digestion.

B. Respiratory System

By first most species have larvae with gill slits and gill arches, but gill filaments develop later. The cobitid *Misgurnus fossilis* has external gill filaments for a time. *Harder* (1954) gave a particularly thorough account for the herring, where the filaments first appear at a body legnth of 20 mm several

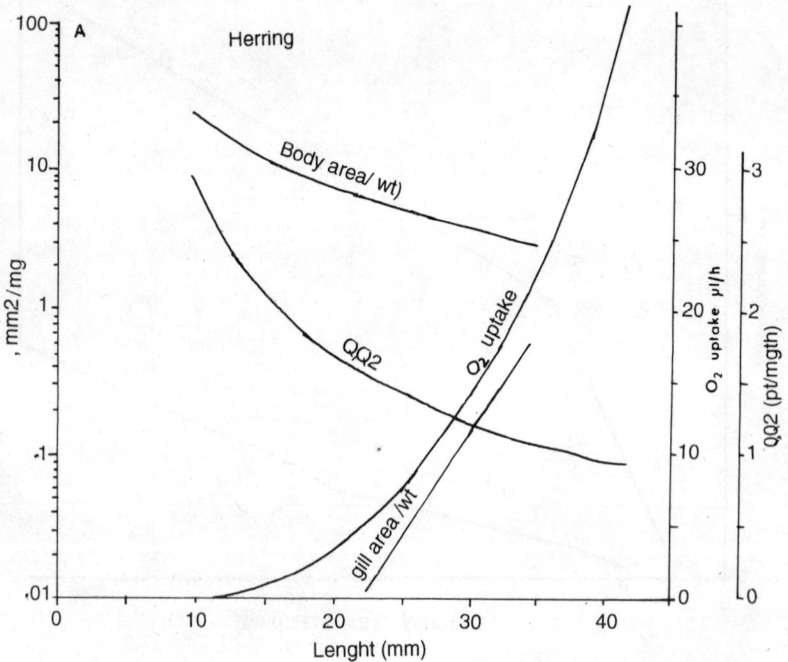

Fig. 2.20. The effect of size on oxygen uptake (ul O_2h), QO_2 (ul O_2/mg dry weight/h), gill area (mm² /mg wet weight) and body area (mm² /mg wet weight) in (A) Atlantic herring *Clupea harengus* and (B) plaice *Pleurovestes platessa* harvae [Data from De Silva (1974).]

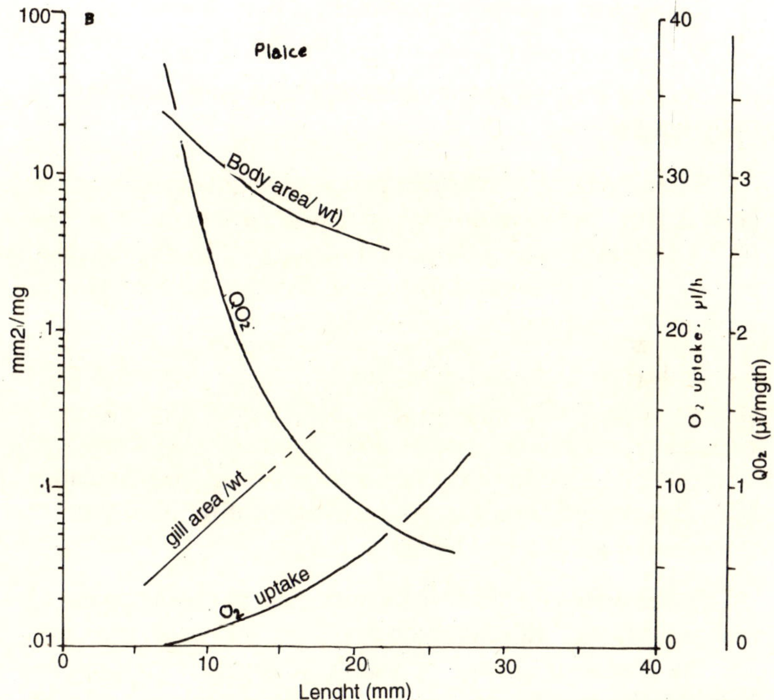

Fig. 2.20 (Continued)

weeks after hatching. *De Silva* (1974) measured the gill area of both herring and plaice larvae during early developmental and related them to the surface area of the body. In the early stages it is clear that respiration is cutaneous. The larval heart is present even before hatching and pumps a colorless body fluid around an as yet unknown vascular system. *Weihs* (1980b) found that yolk-sac larvae of northern anchovy were more active when the oxygen saturation of the water dropped below 60%, suggesting the requirement to renew the ambient water close to the body when oxygen is limiting.

In relating size to respiration the following theoretical conclusions may be drawn :

Total body area for cutaneous respiration α body length^2

Body weight α body length^3

Respiratory area per unit weight α body length^{-1}

O_2 requirement α body weight$^{0.8}$ or body length$^{2.4}$

Specific oxygen requirement (QO_2) α body weight$^{-0.2}$ or body length$^{-0.6}$

Thus the body surfae area per unit weight declines as the body length^{-1}, whereas the oxygen requirement per unit weight (QO_2) declines less rapidly, as body length$^{-0.6}$. This would lead to a critical situation without the development of gills to increase the respiratory surface area.

Although the blood of herring and plaice and many other species does not appear pink until metamorphosis, its precursors, or related substances such as *myoglobin*, can be identified histochemically soon after hatching. The circulating body fluids were reported to be acellular until about 16 mm in the herring and 10 mm length in the plaice.

In the walleye *Stizostedion vitreum* the blood becomes red before hatching, although the development is not precocial. In the precocial larvae of salmonids and elasmobranchs, hemoglobin may be present, and the blood pink, at hatching. In these species the body size is large enough for cutaneous respiration to be inadequate and so increase in oxygen-carrying capacity of the blood is essential, even if the gills are in a rudimentary state.

C. Locomotor System

Most species hatch with V-shaped myotomes acting against the notochord as a *hydrostatic skeleton*. Additional myotomes may be added posteriorly, but the final number is attained during the early larval period. Generally the myotomes, which comprise white muscle, become progressively more complex in shape and interdigitate with adjacent myotomes. The red muscle develops initially as a *myotube*, a superficial cylindrical sheath around the body, e.g., in northern anchovy, zebrafish, herring, and red sea bream *Pagrus major*. In *Coregonus sp.* the red muscle extends as a thin layer dorsally and ventrally from the lateral line. In all these species it later concentrates in a strip in the midlateral position on the flank.

Larvae usually hatct with a primordial medium finfold; the median fins often first appear as a discontinuity in the margin of the finfold, a few fin rays then appear, which gradually increase in number and size. In species with a homocercal tail the caudal fin develops after the tip of the notochord turns up. Lateral fins, used for stability, maneuvering, and sometimes for propulsion, develop differently. Pectoral buds, fin-like structures that lack rays, are often present at hatching. Pelvic buds and fin rays develop later.

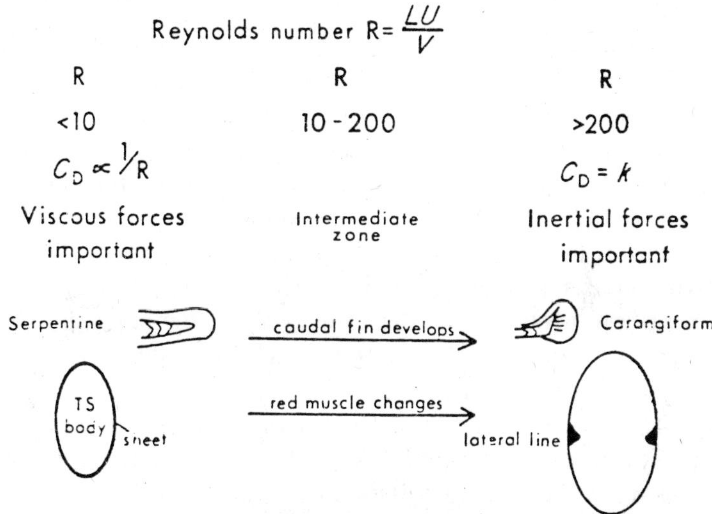

Fig. 2.21. Changes in hydrodynamic considerations as fish larvae grow related to the Reynolds number (R), where *L* is body length, *U* is swimming speed, and *V* is the kinematic viscosity. An inertial regime exists where R > 200 and a viscous regime where R < 10, with an intermediate zone between. The drag coefficient (CD) also changes. At the same time, the larvae change from a serpentine to a carangiform swimming mode as the tail flexion occurs and the red muscle develops from a myotube surrounding the whole body in a thin sheet (TS) to a strip situated along the midlateral position on the flank.

A number of workers have considered the *"locomotor regime"* of larvae of different size. Where the *Reynolds* number (R) is below 10 (in very small larvae), viscous forces are paramount and continuous high-speed swimming is energetically efficient. Where R is greater than 200, intertial forces are more important and beat-and-glide swimming is more efficient. Since R depends on body length and velocity, the hydrodynamic regime changes as the larvae grow or alter their swimming speed. Often they may occupy an intermediate zone between the viscous and inertial regime. During later development there are, in many species, increases in the surface area for propulsion as a result of the appearance of the caudal fin and allometric growth.

Linked to swimming is the buoyancy of larvae. Although unimportant for *demersal larvae*, pelagic larvae can potentially waste much energy in maintaining their position in the water column. As larvae grow and the skeleton ossifies they become heavier and tend more and more to negative buoyancy. Conversely, if they starve, they will tend to neutral buoyancy. The larvae of many species fill the swimbladder soon after hatching, or at the end of the yolk resorption, probably by swallowing air at the surface. This stage, sometimes called *"swim-up"*, can be critical for successful later development, e.g., in the turbot. The time of appearance of the swimbladder varies widely between species. For example, the herring fills its swimbladder at a length of about 30 mm, while the northern anchovy inflates its swimbladder at about 10 mm and the menhanden *Brevoortia tyrannus* at 13 mm. In the physostomatous northern anchovy and menhaden there is a diel rhythm, the larvae filling their swimbladders by swallowing air at the surface at night.

SENSE ORGANS

1. The Eye
In some species the eyes are free of pigment at hatching

and histological examination confirms that they must be nonfunctional. By first feeding the eyes are pigmented, and vision plays a major part in feeding. Of 10 teleost families examined by *Blaxter* and *Staines* (1970), eight had larvae at first feeding with a pure-cone retina, and only an anguillid and a macrourid, caught in deep water, had a pure-rod retina. A pure-cone retina at first feeding was also found in the northern anchovy and in the goldfish. In zebrafish, rods can be identified 9 days after hatching, and at hatching in Pacific salmon; *Oncorhynchus spp.* In the advanced young of the viviparous guppy *Poecilia reticulata*, *Kunz et al.* (1983) found a well-differentiated duplex retina even before birth. Retinomotor movements of the masking pigment and the photoreceptors usually develop concomitantly with the rods so that the process of light-and-dark adaptation is linked to the establishment of a duplex retina.

2. Mechanoreceptors

Free neuromasts are present at hatching in all species examined, usually on the head and sometimes on the trunk. *Disler* (1971) gives a very detailed account of the development of the free neuromasts and lateral line of many freshwater species, including the sturgeon *Acipenser stellatus*, Pacific salmon *O. keta*, and some periods and cyprinids. Generally the initial number of free neuromasts is low but they proliferate and may become regularly arranged along the flank. In marine species similar systems are found in gadids, northern anchovy, Atlantic herring, halibut, plaice and turbot, and spotted bass *Micropterus punctatus*.

The lateral line canals almost invariably develop some time after hatching : at 17 mm in menhaden, 18-20 mm in northern anchovy, 24-26 mm in Atlantic herring, 8 mm in the turbot, 10 mm in the plaice, and 12 mm in spotted bass. Thus young larvae have very incomplete mechanoreceptors. The development of the inner ear is not well-known, except that larvae have one or more pairs of otoliths at hatching, which must give them a basic perception of posture.

Structure and Function

In the majority of cases, the larvae, apart from some highly developed ovoviviparous or viviparous species, or species with very large eggs, so through a massive increase in complexity while free-swimming. Since structures are often absent or incompletely developed, the associated behaviours are also absent or poorly developed. As stated by a number of workers, that *ontogeny is saltatory*, meaning that development proceeds by a series of rather rapid changes in both structure and function with relatively prolonged intervals in between, during which a more-or-less steady state exists as the organism prepares itself for the next rapid change. Thus development does not proceed by a continuous accumulation of small changes as seen in the case of *Abramis ballerus*, a cyprinid. The age and length are related to the development of the sensory, respiratory, digestive, and locomotor systems and associated behaviour and ability to withstand starvation. *Fukuhara* (1985) gives a similarly comprehensive account of the functional morphology of the red sea bream *Pagrus major*.

Allen et al. (1976) related the development of swallowing behaviour, avoidance responses, and shoaling specifically to the development of the pro-otic bullae and swimbladder of Atlantic herring larvae. In a somewhat similar fashion, *Kawamura* and *Ishida* (1985) compared the development of the whole sensory system of the flounder *Paralichthys olivaceus* to primary orientation, feeding, migration, and activity. Considerable differences in morphological and behavioural events can be seen in these two unrelated species; the herring, with an extended larval period metamorphosing into a pelagic schooling species, the flounder, a flatfish with a shorter larval period ending in settlement.

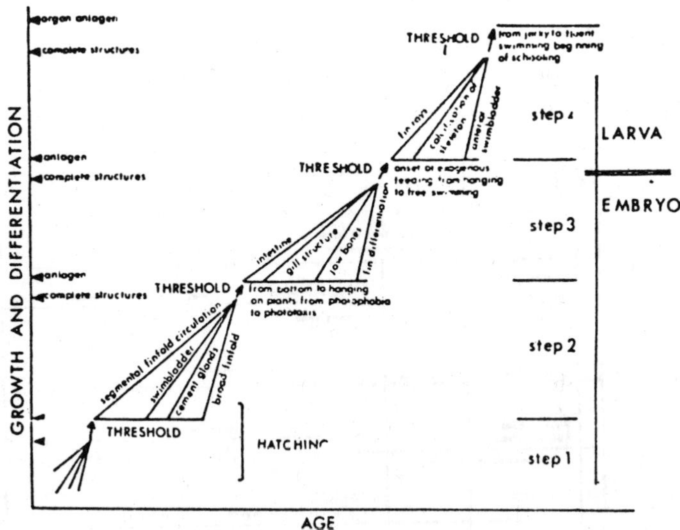

Fig. 2.22. A scheme of consecutive steps in the ontogeny of the Danubian bream *Abramis ballerus*, demonstrating saltatory development. During each step various structures grow and differentiate at different rates but are completed and become functional at the same time, at the end of the step, thus enabling the larvae to make substantial and rapid changes in behaviour. Note according to the author's terminology that the embryo changes to the larva at first feeding, after step 3.

Balon (1980) gives a detailed but similar style of summary of developmental events in five species of charr *Salvelinus*. Despite their close relationship, they show substantial differences in the timing of the appearance of both morphological and behavioural features such as fins, melanophores, branchial respiration and swimbladder filling. The ontogeny of behaviour, especially in salmonids and cichlids, is discussed by *Noakes* (1978), and *Noakes* and *Godin*. These groups are of particular interest because much of the early life history may be passed in gravel beds or under the care of a parent. Such a lifestyle may enhance the protection of the young but it imposes other problems, such as the avoidance of abrasion or cannibalism, which are quite different from free-living fish larvae. Both the behaviourist and the physiologist need to be aware of the ecology and reproductive habits of their experimental material before making evaluations of their data.

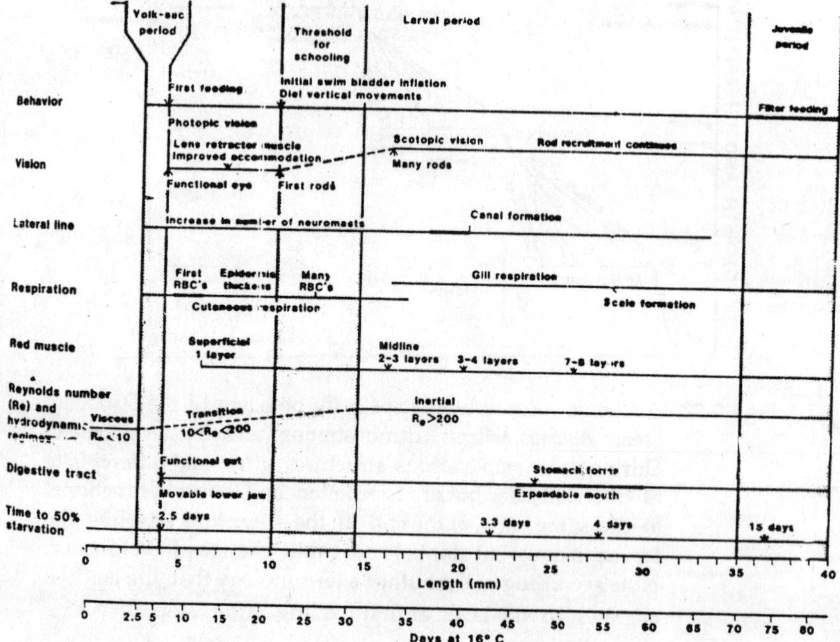

Fig. 2.23. Developmental events during the early life history of the northern anchovy *Engraulis mordax*. RBC, red blood cells; time to 50% starvation is the number of days after which 50% of unfed larvae died (equivalent to point-of-no-return).

CRITICAL PERIODS

The high fecundity of most fishes and the low survival rate of their offspring imply as a high mortality from a number of possible causes: inherited defects, egg quality, starvation, disease, predation. Whether these sources of mortality occur continuously or sporadically or whether there are particularly *"critical periods"* of high mortality is often uncertain. At the

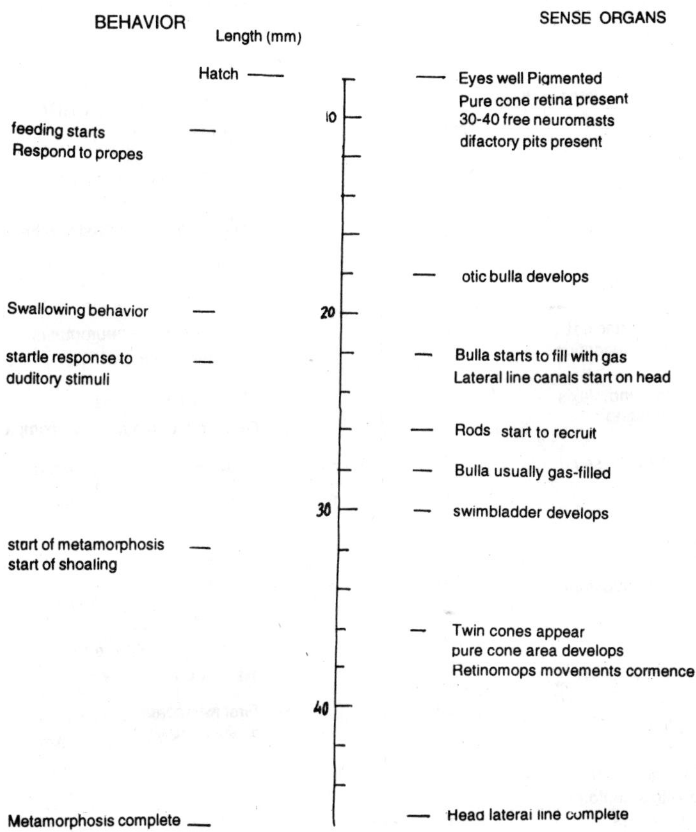

Fig. 2.24. The development of parts of the acoustic system and swimbladder of Atlantic herring *Clupea harengus* in relation to behaviour.

present time it is not possible to conclude the extent to which development is saltatory or gradual. Such a conclusion requires a thorough study of physiology and behaviour as well as anatomy. Because the larvae lack certain behavioural responses, and because they are going through a massive morphogenesis, it seems almost inevitable that critical periods arise through which have to pass to allow development to proceed. These critical periods are especially related to feeding and predation but also to respiration. Some potentially critical periods may be listed as follows :

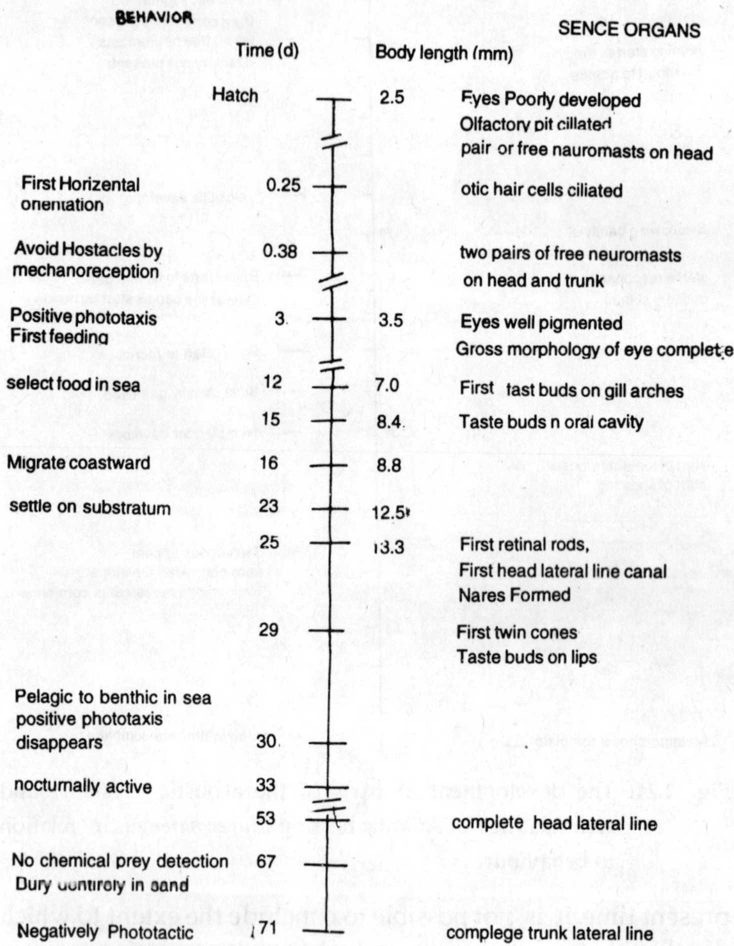

Fig. 2.25. The development of the sense organs and behaviour in the flounder *Paralichthys olivaceus*.

1. Hatching

This depends on the production of hatching enzymes to

break down the tough chorion that protects the embryo from the rigors of wave action or the pressures and abrasion within or on a spawning substratum.

2. First-feeding

Both high mortality at first-feeding in rearing experiments and considerations of brood survival under natural conditions have suggested in the past that one of the main phases of high mortality occurs when the larvae change from endogenous to exogenous sources of food (sometimes called the mixed feeding stage). Obviously relevant to this thesis is the gape of the jaw in relation to the size of prey available and the quantity and quality of the prey. *May* (1974), *Vladimirov* (1975), and *Blaxter* (1984) discuss this problem and conclude that it is likely that mortality often takes place rather steadily over the early life history.   Mortality rates are sometimes as high as in the egg stage of pelagic marine fishes as in the larval stage (5-20% per day), suggesting that predation can predominate over starvation on some occasions. Furthermore, rearing in the absence of predators can lead to excellent rates of survival. First-feeding as the major critical period in development, while sometimes being applicable, should thus be treated with caution as a general concept.

3. Respiration

The disadvantageous decline in body surface area with size in relation to oxygen requirements has already been discussed. There is some evidence of a high phase of mortality associated with early development of the gill filaments, for example, in herring.

4. Swim-up

The first-filling of the swimbladder (and of the bulla system in clupeoids) is essential for maintaining buoyancy and other functions such as hearing associated with a gas-filled swimbladder. In many species, such as sea bass and sea bream, the swimbladder appears soon after hatching.   In the physotomatous salmonids, such as brown trout, rainbow trout, and whitefish *C. clupeaformis*, the swimbladder is filled by swallowing air at the surface at the end of yolk resorption. If access to the surface is prevented, the pneumatic duct remains open and the swimbladder can be filled later. The motivation to fill the swimbladder is very strong; in experiments *Salvelinus (Cristivomer) namaycush* will swim up for at least 270 m without fatigue to fill the swimbladder if it is empty. In physoclists failure to fill the swimbladder can lead to abnormal behaviour and sometimes to death (e.g., in mullet, sea bass, and turbot). In clupeoids avoidance responses to sound stimuli fail to develop unless the bulla contains gas.

5. Metamorphosis

A number of major changes occur in those species where a marked metamorphosis occurs at the end of the larval stage. Increasing conspicuousness, as the transparent larval form is lost, makes it essential for fish to develop other protective mechanisms, such as schooling or burying behaviour, in order to avoid predation. Protective colouration mechanisms also develop as the scales, pigment, and new chromatophores appear. Single fish must be vulnerable at this time, especially in the pelagic habitat, and survival in a schooling species may well depend on early successful aggregation with conspecifics. At or near metamorphosis many marine species move inshore to seek nursery grounds. Flatfishes need to settle on an appropriate substratum such as sand at a near metamorphosis.

3

Yolk Formation

Vitellogenesis is the synthesis of reserve food in the form of yolk. At certain stages of their life histories, the females of egg-laying vertebrates, including most species of fishes, enter a phase of maturation of their oocytes in preparation for ovulation and spawning. Under the mutifaceted influence of hormonal centers such as the hypothalamus and the pituitary gland, the growing follicles synthesize and excrete into the systemic circulation steriod hormones that govern a variety of different metabolic processes. One of the primary target organs for these steriods, particularly *17β-estradiol*, is the liver. The organ, which possesses highly specific binding proteins for *17β-estradiol*, in turn responds to such hormonal stimulus with the synthesis and export of *vitellogenin*. First named by *Pan et al.* (1969), *vitellogenin* constitutes the carrier molecule for various classes of compounds accumulated by the developing oocyte. While the backbone of the *vitellogenin* molecule is a protein chain of substantial size (molecular weight 250,000-600,000), it also carries copious amounts of lipid material, carbohydrate components, phosphate groups, and mineral salts. Following highly selective uptake into the oocyte, the transport molecule *vitellogenin* is broken up and accumulated as egg-specific yolk constituents, such as *phosvitin* and *lipovitellin.*

In addition to these well-known egg components, growing fish oocytes accumulate a variety of other compounds, sometimes in substantial amounts, which play an integral part in the proper development of the embryonic and larval fish. Some compounds serve as a reservoir for energy-demanding processes. In some cases, however, the physiological importance of these compounds is not yet known or can only be inferred from circumstantial evidence. Substances that fall into these categories include glycogen, carotenoids, lectins, sialoglycoproteins, wax esters, and sterol esters.

The early developmental stages of many species of fish entail long periods, sometimes weeks, of starvation prior to their first exogenous feeding. Therefore, the maternal production of *vitellogenin* and the deposition of adequate supplies of yolk, as well as the proper assembly of the oocyte, are essential to subsequent embryonic and larval survival. At this point it should be emphasized that research focusing on piscine systems has been a relatively recent addition to a field that has long been established for amphibians and birds as experimental animals.

VITELLOGENESIS

The hormonal control of exogenous vitellogenesis has been tested experimentally and modified accordingly. Environmental cues such as photoperiod, temperature, feeding, and social factors all regulate the production of gonadotropic hormones (GtH) in the pituitary. In females, the ovaries respond to increased levels of GtH by enhancing estrogen (17β-estradiol and possibly estrone) production and release, into the bloodstream. Estrogens are transported to the possible target tissues bound to specific blood proteins or attached to the ubiquitous blood albumins. They are thought to enter the tissues by facilitated diffusion and in the livers specifically induce the synthesis of *vitellogenin*. Much of the experimental evidence in support of this scheme has been reviewed and can be summarized into several categories:

1. In vitro studies make it possible to identify and characterize highly specific estrogen binding proteins in fish hepatic tissue.

2. The *"natural"* dynamics of hormone titers and vitellogenesis have been studied and indicate that estradiol levels and vitellogenesis are positively correlated and increase in parallel, following photoperiod-related increases in gonadotropins.

3. The effects of experimental treatment of fishes with various hormones, hormonal metabolites, and analogs have been well-documented. Note that in these and many of the studies discussed later, the observed effects are inducible in male and/or immature fish. Injection with pituitary extracts or certain purification fractions induces estrogen synthesis and subsequent vitellogenesis. Injection with estrogen alone or pharmacological doses of androgen induces the appearance of *vitellogenin* in the plasma.

4. In vitro studies of ovaries have confirmed the cells of the follicular epithelium as responsible for the production of *estrogen*. However, certain parts of the fish brain have also been shown to be capable of synthesizing *estrogen*, whereas the quantitative contribution of brain-derived estrogens to circulating estrogens and hence vitellogenesis has not been resolved.

The major components of fish oocytes are derived from the blood-borne high molecular-weight compound *vitellogenin* which is synthesized in the liver of oviparous vertebrates. This supply of oocyte components — especially yolk — from extraovarian sources has been termed *exogenous vitellogenesis*. This classification of *vitellogenin* as a phospholipoglycoprotein already indicates the crucial functional groups that are carried on the protein backbone of the molecule, namely, lipids, some carbohydrates, and phosphate groups. In addition, *vitellogenin* also has strong ion-binding properties and thus may serve as a major supply of minerals to the oocyte.

Hormonal Induction

In the course of the normal events accompanying early vitellogenesis, the follicle cells surrounding the developing oocytes are stimulated to synthesize *estradiol*. *Androgens* constitute the prevailing precursors of *estradiol*, and a slow increase in *estradiol* during the annual cycle indicates early *exogenous vitellogenesis*. After the hormone has entered the target organs, in this case the liver, it binds to highly specific estrogen receptors. In a subsequent step, the receptor/estradiol complex will bind to high affinity sites on the chromatin. While the actual biochemistry of the hormone — receptor complex and the mechanism of its binding to nuclear structures has not yet been entirely elucidated, it has been properly observed that the interaction of the hormone — receptor complex with the DNA leads to modulation of the expression of specific genes. In the case of *estradiol* administration to immature female or male fish, specific activation is directed toward the *vitellogenin gene*, which is located somewhat downstream on the DNA of the actual binding site for the receptor/hormone complex. Thus, the estrogen receptor can be regarded as a gene regulatory protein.

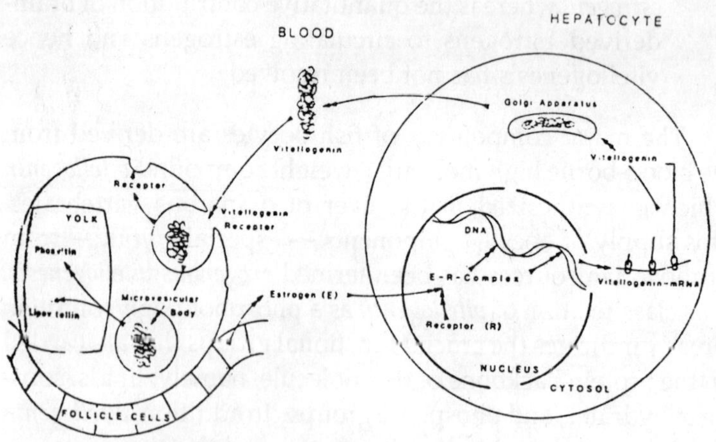

Fig. 3.1. Simplified feedback system between ovary and liver during exogenous vitellogenesis.

The conversion of *androgens* into *estrogens* is catalyzed by the enzyme *aromatase* and occurs not only in the ovary but also in the brain of all vertebrates; *aromatase* activity is exceptionally high in teleosts. However, the *estrogens* thus produced in the fish brain apparently is not released into the circulation, unless chemically altered, which results in its ultimate removal from the circulation by gill tissue. Exertion of the multifaceted biological effects of this hormone will consequently be restricted to the local brain level and therefore cannot be implicated in the initiation of *exogenous vitellogenesis*.

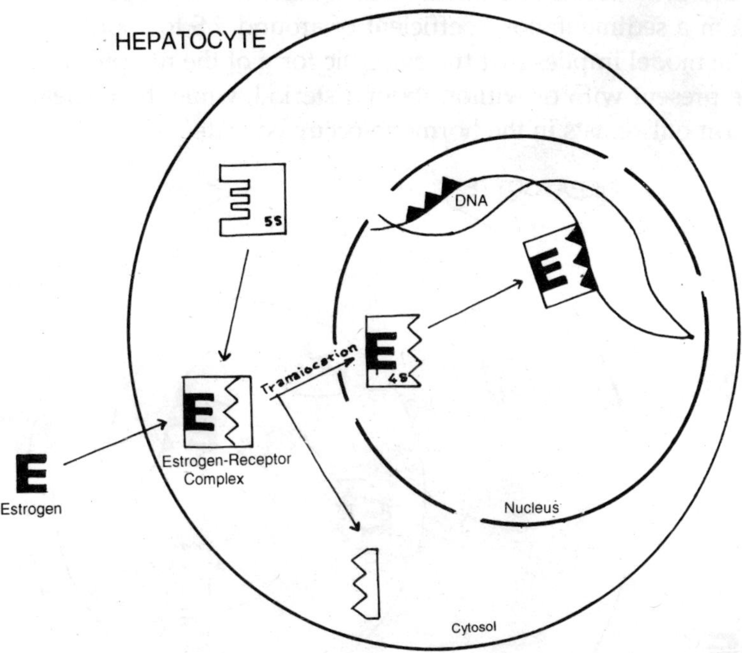

Fig. 3.2. Two-step model of estrogen-receptor mechanism. Estrogen binds to cytosolic 5-S receptor proteins in the target cell (liver). During translocation from the cytosol to the nuclear matrix, the estrogen-occupied receptor (estrogen-receptor complex) changes from 5 S to 4 S and acquires high affinity for the chromatin. The estrogen-receptor complex builds to specific sites on the DNA and among other things leads to activation of the gene coding for vitellogenin.

Estrogen Receptors in Target Cells

The prevailing model for the cellular distribution and action of steriod hormone receptors has been the so-called *"two-step model,"* which was first proposed by *Gorski, Jensen,* and co-workers. According to this model *estradiol* — or any other steriod hormone for that matter — enters the target cell by diffusion and binds to a specific cytosolic form of the receptor protein. Subsequently, the estrogen-receptor complex undergoes a transformation from a non-DNA binding form to a species that does and is translocated to the nucleus. Receptor transformation is also reflected in a decrease in receptor size from a sedimentation coefficient of around 5 S to about 4 S. The model implies that the cytosolic form of the receptor can be present with or without bound steriod, while the nuclear form only exists in the hormone-occupied state.

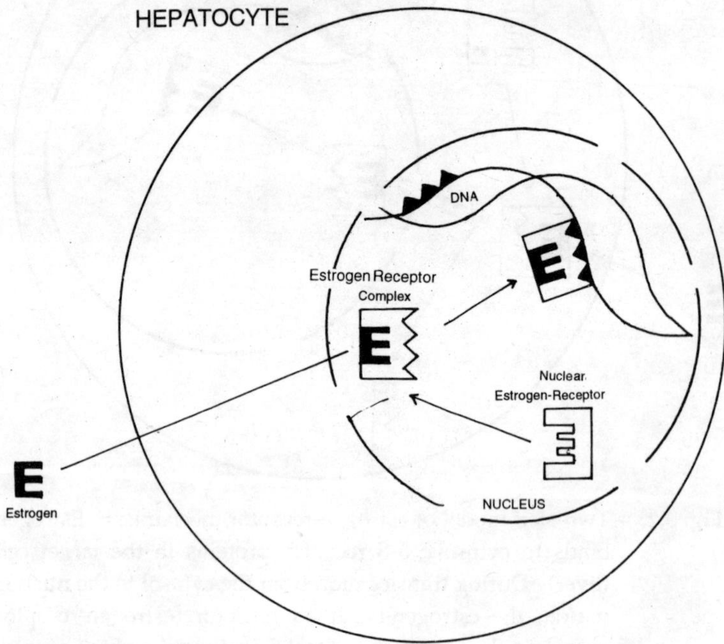

Fig. 3.3.   Nuclear model of estrogen-receptor mechanism.   Estrogen enters the target cell and binds to highly specific nuclear

proteins. Estrogen-receptor complex displays high affinity for chromatin, binds to specific sites on the DNA, and, among other things, leads to activation of the vitellogenin gene.

Recently, a different model of steriod receptor localization has been proposed, disbanding the existence of the cytosolic receptor. Using powerful immunocytochemical techniques, the proponents of this *"nuclear model"* localized all receptor molecules, unoccupied or occupied, in the nuclear region. In this model, no nuclear translocation of receptor/hormone complex occurs and the receptor attain its increase in affinity for nuclear structures by binding hormone directly within the nuclear compartment. Highlighting the merits of the comparative approach, the work of *Callard* and *Mak* (1985) on the estrogen receptors of the testes in an elasmobranch fish *(Squalus acanthias)* support this novel working model; both occupied and unoccupied receptors are exclusively associated with the nuclear compartment.

In recent years, evidence has accumulated that many steriod receptors fall into an interesting group of proteins that are regulated by reversible covalent modifications through phosphorylation/dephosphorylation. Experimental data suggest that phosphorylation is a pre-requisite for the hormone binding activity of the receptor protein and that dephosphorylation leads to an inactivation of the receptor molecule. While most proteins, including the chicken progesterone receptor and the glucocorticoid receptor of fibro-blasts, are phosphorylated at serine residues within the polypeptide chain, the calf uterine *estrogen* receptor protein is phosphorylated on tyrosine, a relatively unusual phosphorylation site for covalent modification.

In light of the present controversy over the actual cellular localization of unoccupied estrogen receptor, where the evidence points to the cytosolic receptor as a possible experimental artifact, it is somewhat confusing to reconcile the new interpretation with the fact that phosphorylation of the calf estrogen receptor is brought about by a specific, calcium- and calmodulin-dependent cytosolic receptor kinase. Not surprisingly, however, the dephosphorylation-dependent inactivation of the receptor is caused by a phosphatase with exclusive localization within the nuclear compartment.

While the validity of the existing models has not been tested for hepatic tissue in piscine systems, the presence of highly specific estrogen receptors has been verified for a number of different species of fish, namely, *Gobius niger,* Pacific hagfish *(Eptatretus stouti;* Atlantic salman *(Salmo salar, sea-raven (Hemitripterus americanus),* winter flounder *(Pseudopleuronectes americanus),* scuplin *(Myoxocephalus octodecimspinosus;)* rainbow trout *(S. gairdneri)* and most recently in the brown trout *(s. trutta).*

The livers of nonvitellogenic or male Atlantic salmon *(S. salar)* contain specific high-affinity estrogen binding proteins in the cytosolic fraction, while the liver nuclei reveal low concentrations of high-affinity estradiol-binding components. Injection of pharmacological doses of estradiol in the fish leads to a transient depletion of the cytosol binder and the appearance of exceptionally high concentrations of estradiol binding sites in nuclear salt extracts. In naive fish, the concentration of binding sites in the cytosol was 640 fmol/g liver, while their concentration in liver nuclei was 150 fmol/g liver. After the administration of a single dose of estradiol, the concentration of nuclear binding sites increases almost 80-fold to above 11 pmol/g liver. The induction is apparently highly specific for estradiol and, after a single injection of estradiol into an experimental fish, reaches a maximum of receptor response after some 24 h. Such response can just as easily be elicited in cultured hepatocytes, using physiological, rather than the more usual pharmacological, doses of hormone.

Most enzymes involved in liver cell metabolism, for instance, occur in concentrations ranging from 10 to 100 nmol per gram of liver, that is, at least two orders of magnitude higher than the maximum amount of estrogen receptor. One gram of liver contains about 0.9 µg estrogen receptor, assuming a molecular mass of about 80 kDa, compared with almost 500 µg pyruvate kinase alone.

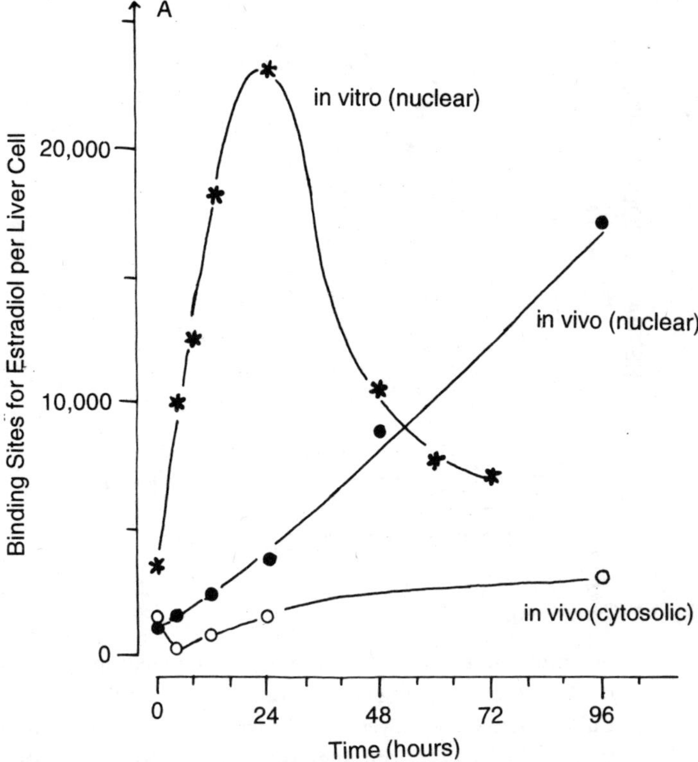

Fig. 3.4. (A) Temporal changes in hepatic estrogen receptor after estrogen administration. Fish (O●) or isolated hepatocytes in culture were treated with a single dose of estradiol [5 mg (kg live weight)[-1] and 1 µm for cultured cells, respectively]. Estrogen binding sites were determined after the indicated time in nuclear (●,*) and cytosolic (O) fractions. Numbers of assayable binding sites in the cytosolic fraction are temporarily depressed, while numbers of nuclear binding sites continue to increase until the end of the experimental period (120 h). In cultured liver cells, a maximum of response is reached about 24 h after the addition of estradiol. Vitellogenin can first be detected

Fig. cont....

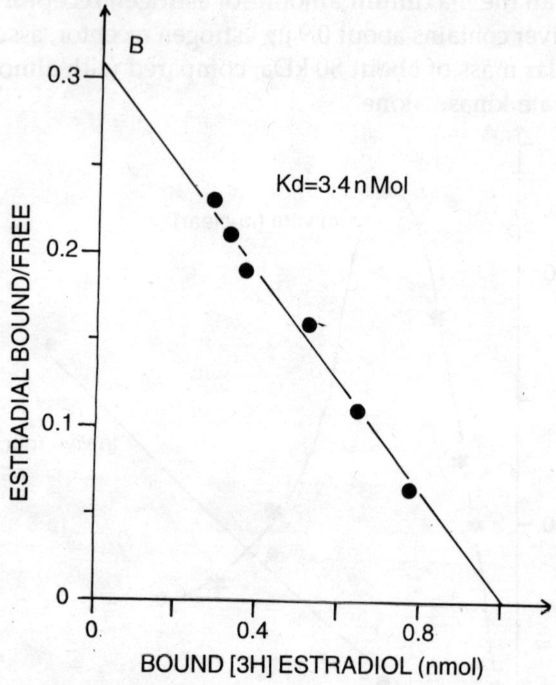

Fig. 3.4 (Continued)

by immunoprecipitation in the hepatocyte medium after 24 h. (B) Estrogen-binding characteristics of salmon liver nuclei. Cultural hepatocytes were treated with estradiol as indicated in (A). At 12 h after hormone exposure, cells were harvested and nuclear extracts were analyzed for receptor activity and characteristics (Scatchard analysis). The dissociation constant (Kd) for estradiol is computed by linear regression and found to be 3.4 nM^{-1}. The dissociation constant for highly specific estrogen binders from liver nuclei in estrogen primed fish is between 5.4 and 5.9 nM.

A similar increase in the total level of cellular estrogen-binding sites, where the increase in nuclear binding sites more than offsets the decrease in cytosolic sites, is observed when cultured salmon liver cells are exposed to physiological doses of estradiol. The time course and the peak concentration reached differ somewhat between the in vivo and the in vitro systems, possibly due to the amounts of estradiol administered and enhanced metabolism in the isolated liver cells. While this particular subject was not the focus of any of the studies on fish, it is interesting to note that the metabolism of the steriod hormone itself and not so much the kinetics of the hormone — receptor decay influences the effectiveness of the hormone administration. In tissue cultures of *Xenopus* hepatocytes, for instance, estradiol turnover is rather rapid, especially in cells derived from male toads. Here, the half-life of the molecule is in the range of only 40 min at 20°C. It is obvious that if similar conditions are found in isolated fish cells, very little exposure is needed to initiate the transcription of estradiol-dependent genes, in this case the estrogen-receptor gene, thus setting up an interesting positive feedback system.

Compared with other vitellogenic vertebrates, the teleost liver appears to be the richest source of highly specific estrogen receptors, making the fish liver an ideal model system to study the induction of receptor and analyze in detail the mechanism of hormone — receptor and receptor-chromatin interactions in lower vertebrates. The values listed in Table 3.1 compare the magnitude of the receptor induction in *S. salar* with other oviparous vertebrates utilized in the analysis of the biochemistry of the estrogen receptor. Recently, it was found that the salmon is not unique in this respect, nor is such largesse of response restricted to salmonid fish. Liver nuclei isolated from the sea-raven *(H. americanus)*, the longhorn sculpin *(M. octodecimspinosus)*, winter flounder *(P. americanus)*, and the rainbow *(S. gairdneri)* all accumulate similarly high concentrations of estrogen binding proteins as the Atlantic salmon. In a goby *(Gobius niger)*, an elasmobranch *(Potamotrygon, ssp.)*, and a hagfish *(Eptatretus stouti)*, on the other hand, numbers of estrogen-binding sites in the liver are more than an order of magnitude lower than in the salmon.

TABLE 3.1

Magnitude of Estrogen-Receptor Response in Vitellogenic Vertebrates

	Naive animals	*After induction with estradiol*
Salmo salar	0.15	> 12
Xenopus laevis	0.2	2.5
Gallus domesticus	0.1	0.4

While the actual source of these comparatively large amounts of the gene-regulatory estrogen receptor in the piscine system has not been elucidated yet, some indirect evidence in other vitellogenic vertebrates indicates that de novo synthesis of receptor protein is involved. Again, the highly sensitive piscine system seems ideally suited to supply mechanistic insight into the source of *"induced"* receptor molecules by molecular biology techniques, similar to the ones used to assess the transcriptional activity and mRNA longevity of the albumin gene in *Xenopus* for instance.

The fish receptor proteins resemble the receptors from other vertebrates in many respects. The salmon receptors are characterized by a high specificity for *extradiol* and they do not bind *progesterone, hydrocortisone,* or *dihydrotestosterone.*   In agreement with studies of receptors from many other vertebrate sources, the fish receptors display high affinity for the nonsteroidal estrogen diethylstilbestrol as well as for the nonsteroidal antiestrogen 4-hydroxytamoxifen. The estrogen binding proteins in the liver of the hagfish *E. stouti,* on the other hand, display unique features, different from the global vertebrate picture. This species possesses nuclear estrogen receptors with a lower affinity for extradiol than other vertebrate counterparts. However, the hagfish system is unusual in that *estrone* or *estriol* displaced *estradiol* from the nuclear binding components as efficiently as estradiol or diethylstilbestrol. In other vertebrates, binding affinities for *estriol* or estrone are usually more than an order of magnitude lower than for *estradiol.*

In the rainbow trout, *estrone* administration leads to the induction of *vitellogenin* synthesis in the liver and its release into the bloodstream, but *estrone* displays only 5% to 12% of the potency of estradiol. It seems that one of the functions of *estrone* in vivo may be to prime hepatic tissue for subsequent exposure to *estradiol* and thus to potentiate the vitellogenic response of the hepatocytes to estradiol.

In the annual cycle of the rainbow trout, the liver is exposed to differing concentrations and ratios of *estradiol* and *estrone*. While blood concentrations of both *estrogens* increase during early vitellogenesis, the first phase of vitellogenesis is dominated by *estrone*, which increases by a factor of 10 altogether. During the later stages of exogenous vitellogenesis, *estradiol* reaches blood concentrations of 60 ng/ml, reflecting an increase of 60-fold. It would be interesting to analyze whether similar changes are reflected in the abundance or preference of specific *estrogen* receptors in the nuclei of the liver, the main target organ for ovarian *estrogens*.

In addition to the *estrogens*, *androgens* are able to elicit a vitellogenic response in teleost fish, albeit only when administered in pharmacological doses. Interestingly, at least in G. *niger* this response appears to be mediated by *androgen* binding to the estrogen receptor rather than through the nuclear androgen receptor itself. Similarly, high doses of *androgen* fed to juvenile salmon may lead to a pronounced feminization of some fishes, although the molecular mechanisms for these phenomena remain to be analyzed. The sedimentation coefficient of the salmon receptors protein of 3.6 S indicates that it may be a little smaller than the nuclear estrogen receptors of birds or mammals, but it falls into the same range as the receptor isolated from *Xenopus laevis*.

Plasma Binding Proteins
The ovarian cells synthesize the steriod. The steriod hormones produced and released by the ovarian cells hormones and then transport of their target tissue in the systemic

circulation. Although probably not a target issue in itself, fish plasma displays a certain degree of steriod-binding capacity. For a variety of fishes as well as other vertebrate groups, such *"sex-steriod binding proteins"* have been characterized numerous times. Since steriod hormones exert their biological functions only in the free and not in the bound form, such plasma steroid-binding proteins may serve to buffer free steroid concentrations in conditions of high steriod turnover, thus obviating time consuming de novo synthesis.

In the plasma of *S. salar*, two differing *estradiol* binding components are abundant, one with high affinity and one with low affinity for *estradiol*. In contrast to the highly estradiol-specific nuclear receptors inducible in the salmon liver, neither of the plasma-binding components are competed for by the nonsteroidal estrogen diethylstilbestrol. Furthermore, again differing from the situation in the liver, the antiestrogen 4-hydroxytamoxifen does not compete with *estradiol* for binding to the high-affinity estrogen binder in plasma. Experiments also indicate that the androgen dihydrotestosterone as well as *progesterone* and *estrone* reveal considerable affinity for the plasma binder and are likely to compete with *extradiol*.

Hepatic Events
When the binding of the estrogen/receptor complex to the nuclear DNA is delayed for few hours, a variety of changes in liver cells are initiated that are consistent with a substantial increase in the capacity for protein synthesis and export — plasma concentration of vitellogenin may reach 50 mg/ml. Indeed, naturally vitellogenic fish reveal much higher rates of hepatic protein synthesis than nonvitellogenic fish, a phenomenon that can be provoked by *estrogen* administration in vivo as well as in vitro. Several ultrastructural differences are observed between liver cells from immature and vitellogenic fish. In the red grouper *(Epinephelus akaara)*, vitellogenic livers are characterized by expanded nuclear envelope cisternae, swollen mitochondria, and much enhanced rough endoplasmic reticulum, Golgi apparatus and secretory vesicles. Several

studies indicate an increase in hepatosomatic index, and, at least in the red grouper, this appears to be due to a rise in cell lipid and water content rather than proliferation of cell numbers. In the Atlantic salmon, as in the flounder, estradiol administration leads to increases in liver protein, total RNA, and total nuclear count. Since at the same time the liver volume and weight increase, calculated on a unit weight basis, only the amount of cellular RNA is augmented significantly. It can be concluded that in these two species of fish, differing from the grouper, hyperplasia rather than hypertrophy appears to be responsible for the enhanced liver weight. The more than 30% increase in cellular RNA content (on a unit weight basis) is yet another indication of the increased biosynthetic activity of the liver where the de novo synthesis of messenger RNA for vitellogenin may account for some of the observed increase in total RNA.

A larger proportion of the newly synthesized RNA in hepatic tissue following the exposure to *estradiol* is due to apparent increases in the amounts of ribosomal RNA, which can be explained by the massive increases in rough endoplasmatic reticulum observable in liver micro-graphs of fish and other oviparous vertebrates. Since *estrogen* administration is responsible for the proliferation of cell structures, such as endoplasmic reticulum (ER), Golgi vesicles (and turnover), and mitochondria, genes coding for any of these structures must have been activated or estrogen administration must have at least led to increased translational activity involving existing mRNAs.

All hepatocytes are metabolically identical. It has been shown rather conclusively that rat hepatocytes in the perivenous and periportal regions of the liver possess differiong metabolic functions, with anabolic pathways such as gluconeogenesis, fat synthesis, and proteins synthesis being favoured in the better oxygenated periportal cells. *Estrogen* treatment of fish also appears to result in a general gearing up to metabolism to provide the large amounts of energy and reducing power (NADPH) necessary for protein and lipid synthesis. *Ng et al.* (1984) further report significant and substantial increases in transaminases and enzymes of the Krebs cycle and glycolysis. On the other hand, naturally vitellogenic female sockeye salmon (*Oncorhynchus nerka*) on their spawning migration do not increase any specific metabolic machinery in liver, apart from the general augmentation due to an—probably estradiol-dependent — increase in liver weight.

In vivo treatment of male flounder (*Platichthys flesus*) results in an increase of protein synthetic activity when assessed in an in vitro systems. That such stimulation may be a direct effect of estrogen or liver cells was recently demonstrated in hepatocytes isolated from juvenile coho salmon (*O. kisutch*. Hepatocytes treated with *17β-estradiol* and exposed to either [^{14}C]serine or [^{14}C]glycine exhibit an increase in radioactivity precipitable by trichloroacetic acid (TCA), a decrease in TCA-soluble radioactivity, and enhanced release to TCA-precipitable radioactivity into the medium compared with untreated controls. Originally it was observed that liver slices, from cod (*G. morhua*) treated with estradiol incorporated labeled [^{14}C]leucine into "egg proteins." All this evidence confirmed that in the teleost fishes, just as in other oviparous vertebrates, estradiol leads to the rapid and specific hepatic synthesis of the egg-yolk precursoe vitellogenin.

TABLE 3.2

Cellular Events Associated with the Estrogen-Dependent Induction of Vitellogenesis in Teleost Hepatic Tissue

Transient decrease in cytosolic estrogen receptor protein

Induction of nuclear estrogen receptor protein

Increase in hepatosomatic index due to hyperplasia or hypertrophy

Proliferation of Golgi apparatus

Increase in cisternae of the nuclear envelope

Synthesis of ribosomes

Polysome assembly

Increase in rough endoplasmatic reticulum

Swelling of mitochondria

Appearance of a new species of mRNA (vitellogenin)

Increase in protein synthetic activity

Synthesis of vitellogenin

Increase in cellular RNA

Increase in lipid metabolism (?)

Augmented output of very low density lipoproteins (VLDL)

Decrease in glycogen content per cell

Increase in metabolic enzymes

Higher amount of hepatic DNA

Increase in hepatic water content

Vitellogenin

The *Vitellogenin* molecules have been isolated from a number of different fishes. Interestingly, the fishes display a much higher variability in the different parameters, such as molecular weight, degree of phosphorylation, degree of lipidation, or subunit composition than their amphibian or avian counterparts. As the example of the tetrameric vitellogenin in the Japanese eel shows not all fishes *vitellogenins* are dimers — for instance, and in the case of the brown bullhead, the messenger RNA for vigellogenin is substantially smaller than that for any other vertebrate. Even within the same species, in this case the rainbow trout *(S. gairdneri)*, a large variation in the apparent molecular weight of the vitellogenin is noticed, which may be due to different methodologies used, different degrees of proteolytic breakdown, or dephosphorylation occurring during the isolation. Some degree of heterogeneity in vitellogenin may be due to the fact that in the fishes, as in other vertebrates, vitellogenin is not encoded by a single gene, but rather by a family of slightly different genes.

TABLE 3.3

Molecular Weights of Native Vitellogenin and Subunits from Fish Other Than Rainbow Trust

Species	Native molecular mass (kDa)	Subunit molecular mass (kDa)	Method[a]
Carassius auratus (goldfish)	326 380	140-156 140-147	Native/SDS-PAGE Native/SDS-PAGE
Gadus morhua (cod)	400	—	Gel filtration
Anguilla japonica (Japanese eel)	350	85	Gel filtration and SDS-PAGE
Fundulus heteroclitus (killfish)	—	200	SDS-PAGE
Ameiurus nebulosus (brown bullhead)	—	145	b
Platichthys flesus (flounder)	550	—	Gel filtration
Oncorhynchus kisutch (coho salmon)	390[c]	—	Gel filtration
Salmo salar (Atlantic salmon)	495 and 520	-	Gel filtration
Salmo trutta (brown trout)	440	—	Gel filtration
Heteropneustes fossilis	550	—	Gel filtration

[a] SDS-PAGE: sodium dodecyl sulfate polyacrylamide gel electrophoresis.

[b] Translation product of the prevailing mRNA induced by estradiol treatment of bullhead catfish.

[c] Lipovitellin from coho salmon eggs.

As long as the translation of one isolated *vitellogenin* messenger RNA (mRNA) in his salmonid fish leads to different molecular-weight estimates for the *vitellogenin* monomer, more attention will have to be devoted to multiple *vitellogenin* genes, to strain differences within one species, or to possible partial degradation of this large molecule or its mRNA — which do not critically affect the immunological reactivity — before a definite answer with respect to molecular weight and phosphorylation sites can be given.

Estradiol treatment leads to the appearance of a specific high-molecular-weight species of messenger RNA in the rainbow trout. Using cytoplasmic polyadenylated RNA isolated from these estradiol-exposed trout in a cell-free translation system, *Chen* (1983) was able to synthesize a 160,000-Da polypeptide that was chemically, immunologically, and electrophoretically identical to the authentic vitellogenin monomer. Similar results were obtained for the same species by *Valotaire* and co-workers (1984), although their larger mRNA (7200 nucleotides) upon translation yielded a considerably larger (200,000 Da) polypeptide, which was immunoprecipitable with antibodies against trout serum *vitellogenin*.

Posttranslational Modifications

The biochemical information concerning *vitellogenin* clearly indicates that a great deal of posttranslational modification must occur in the liver cell to reach the finished product seen in the serum. First, the protein backbone of the *vitellogenin* is synthesized on membrane-bound ribosomes, a feature that it shares with other proteins destined to be secreted from the hepatocyte. In subsequent steps, the molecule must be lipidated, glycosylated, and phosphorylated. It has been suggested that all these processes occur on the membranes of the endoplasmatic reticulum and that they are already initiated while the polypeptide chain is being translated, although this view has been debated by *Gottlieb* and *Wallace* (1982). Finally, existing "pro" sequences or signal peptides have to be removed before *vitellogenin* is packaged into Golgi vesicles and secreted into the bloodstream.

Fish vitellogenins carry a certain number of phosphate groups, some of it as protein phosphorus, in a region that in the mature oocytes becomes deposited as phosvitin. Generally, the molecule is phosphorylated on serine moieties, and since the degree of phosphorylation of delipidated piscine *vitellogenins* ranges around 0.6-0.7% (by weight) (i.e., only about 50% of the protein phosphate content in other vertebrates), the serine content must be comparatively lower. Experimentally, this alkaline-labile protein phosphorus, which is specific to naturally or induced vitellogenic animals, has been utilized repeatedly for the determination of the degree of the vitellogenic response in fishes. Vitellogenic female fishes contain between 20 and 100 µg of protein phosphorus per milliliter of plasma, while untreated males contain less than 5 µg/ml. In spite of the large amounts of protein phosphate moved through the plasma compartment during vitellogenesis, in the unfertilized egg inorganic phosphate, and phospholipid, and not protein-bound phosphate, make up the bulk of the nuclear magnetic resonance (NMR) visible [^{31}P]phosphate. This observation indicates that additional maternal sources must supply phosphate and phospholipids to the oocyte. Also substantial dephosphorylation of vitellogenin-derived phosphoproteins during transmit through the oocyte or following deposition in the yolk may explain the low protein phosphate content of mature oocytes. The highly charged phosphate component also gives the vitellogenin molecule its high ion-binding capacity. Teleost vitellogenins are known to bind ions such as calcium, magnesium, or ion efficiently and thus may designate an important vehicle for mineral supply to the growing oocyte. In fact, the competition of vitellogenin with chelating substances has been used successfully to isolate fish vitellogenins from other plasma proteins.

In contrast to the phosphate content of fish *vitellogenins,* which is lower than that of other oviparous vertebrates, the amounts of lipid material carried on the *vitellogenin* molecule are generally about twice as high as for other vertebrate groups. The lipid content of *vitellogenin* ranges around 20% by weight in fishes as different in lifestyle and feeding preferences as the goldfish 21%; rainbow trout 21%; 18%, sea-trout 19%; or the elasmobranch dogfish 18%. The bulk of this lipid material, which later forms the lipovitellin moiety of the yolk, can be classified as polar lipid. In rainbow trout vitellogenin, for instance, polar lipids make up some 82% of the total. Generally the mature oocyte, however, contains much larger percentages of triglyceride, and it is therefore reasonable to assume that sources other than vitellogenin must supply the oocyte with nonpolar lipids, such as triglycerides, sterylesters, sterols, and wax esters. In this context it is interesting to note that dietary manipulation of free fatty acids in trout is reflected in altered fatty acid composition of serum lipids, but not of the lipoproteins, which are most important during vitellogenesis.

While fish *vitellogenin* is known to contain carbohydrate groups, little concrete information on the amount, nature, and linkages of the carbohydrates is available. However, it is known that for many proteins, successful glycosylation is a prerequisite for excretion of the export protein. In other cases, such as the chicken ovalbumin, which usually occurs in glycosylated form, no glycosylation is required for excretion. Experiments utilizing tunicamycin, a specific inhibitor of N-glycosylation, revealed that the absence of the oligosaccharide side chains from the ovalbumin molecule had no effect on its secretion. For comparative purposes and from an evolutionary perspective, it would be a rewarding task to determine the group of glycoproteins to which the fish vitellogenins belong and whether successful glycosylation is a prerequisite for excretion from the hepatocyte. Obviously, there is a large information gap between the process of glycosylation of the vitellogenin molecule in the liver and the presence of large amounts of sialoglycoproteins in fish eggs.

Vitellogenin could be detected in the blood but not in the livers of estradiol-treated rainbow trout, a result that was first interpreted to indicate that vitellogenin is rapidly secreted following synthesis. However, *Nunomora et al.* (1983), using the peroxidase-antiperoxidase complex method (immunologically specific for vitellogenin), were able to localize significant amounts of vitellogenin in livers of estradiol treated rainbow trout (Salmo gairdneri), chum salmon *(O. keta)*, or charr *(Salvelinus leucomaenis)*. Similarly, So and co-workers (1985) detected cross-reactivity of antibodies against salmon *(Salmo salar)* vitellogenin with liver extracts of vitellogenic fish. These results can be reconciled by the fact that van *Bohemen et al.* (1982b) assayed for vitellogenin by molecular weight determination on sodium dodecyl sulfate (SDS) polyacrylamide gels and thus screened for nature vitellogenin rather than immunoreactive components.

More information regarding posttranslational modification is available from other vertebrate systems. Recently, rooster hepatocytes were used to determine the probable sequence of events in hepatic vitellogenesis. Precursors (pVTG I and pVTG II) for each of the two types of avian vitellogenin (VTG I and VTG II) were found in hepatocytes of roosters treated with estrogen by pulse-labeling with [^{3}H] serine and pulse-chase experiments. The molecular weights of the precursors were lower than those of the mature vitellogenins as determined by SDS gel electrophoresis. However, further analysis of these polypeptides by immunological methods, peptide mapping, and molecular weight determinations by gel chromatography revealed that the precursors are similar to mature vitellogenin in size and degree of glycosylation, but are not phosphorylated. *Wang* and *Williams* (1982) could also show that highly phosphorylated proteins, such as mature avian vitellogenin, will yield erroneously high molecular weights on SDS gels. The very small quantities of phosphorylated vitellogenin inside of these hepatocytes led *Wang* and *Williams* (1982) to suggest that phosphorylation is rapidly followed by secretion. Their determination of vitellogenin molecular weight of gel chromatography caused the same authors to revise the "accepted" molecular weight for the avian vitellogenin monomer from above 235,000 to 180,000. In light of the obvious controversies about the actual molecular weights of piscine vitellogenins, even within the same species, this type of multifaceted approach represents a fertile area for research on vitellogenesis in fish. In this field, the isolated hepatocyte systems would appear to be an excellent, as yet underutilized, experimental tool. Recently, several laboratories have been able to prove that fish hepatocytes in suspension or in primary culture are highly responsive to estrogen and that hepatocytes isolated from primed fish will synthesize and excrete large amounts of vitellogenin in vitro.

From the reviewed studies on fishes and other egg-laying vertebrates, a preliminary picture of the sequence of events implicated in exogenous vitellogenesis can be synthesized. Unfortunately, especially for the fishes, many parts of the scheme require a major concerted research effort from biochemists, physiologists, and molecular biologists alike to replace speculation and add information on actual mechanisms. The major task would be to successfully utilize the vast potential of piscine system to elucidate and understand the estrogen receptor mechanism, its interaction with the nuclear DNA, and not least the subsequent gene activation. Further challenging topics include the diverse posttranslational modifications of the vitellogenin molecule occurring in the liver cell. Furthermore, the particular intracellular structures where the individual steps occur have to date eluded identification.

Recent work on the *number* of similar *vitellogenin* molecules of *Xenopus* has revealed that the situation is not quite as clear-cut or simple as it first appeared. Rather than being just one protein, coded for by one type of mRNa, a whole family of vitellogenin genes is in existence, all of which give rise to slightly different vitellogenin molecules. These in turn supply the growing oocyte with the different building blocks for at least five different types of yolk polypeptides, namely lipovitellins 1 and 2, phosvitin, and phosvettes 1 and 2, which in themselves are somewhat heterogeneous. Lipovitellins 1 and 2 can each be resolved into three differing polypeptide components, while dephosphorylated phosvitin yields two polypeptide bands of different molecular weights on SDS electrophoresis. Phosvettes are relatively small phosphorylated components with single polypeptide chains. From their study on the vitellogenin of *Xenopus*, *Wiley* and *Wallace* came to the conclusion that the whole gamut of yolk proteins is derived from multiple vitellogenin molecules and also that the phosvettes are alternate cleavage products from homologous regions of different parent vitellogenins. Using complimentary DNA copies of *Xenopus* vitellogenin mRNA, *Wahli et al.* (1981) deduced that vitellogenin is encoded in a family of at least four expressed genes.

TABLE 3.4

Suggested Sequence of Events during Hepatic Synthesis of Vitellogenin

---

Nuclear compartment
   Activation of vitellogenin gene through binding of receptor-hormone complex to specific regions of the nuclear DNA
   Transcription and presence of primary transcript in the nuclear compartment
   Processing of primary transcript
   Translocation to cytoplasm
Rough endoplasmatic reticulum
   Polysome assembly
   Translation of vitellogenin mRNA
   Processing of previtellogenin subunits
   Phosphorylation at serine residues[b]
   Lipidation[c]
   Translocation to smooth endoplasmatic reticulum
Smooth endoplasmatic reticulum
   Further phosphorylation at serine residues[b]
   Translocation to Golgi apparatus
Golgi apparatus
   Glycosylation
      Mannose
      N-Acetylgulcosamine
      N-Acetylmeuraminic acid, etc.
   Lipidation[c]
   Removal of existing signal peptides
   Dimerization
   Phosphorylation at serine residues[d]
   Excretion into systemic circulation

---

[b]In Xenopus, phosphorylation occurs in the rough endoplasmatic reticulum and the smooth endoplasmatic reticulum only.

[c]The exact cellular site of the noncovalent attachment of lipid to vitellogenin is still under debate.

[d]In the chicken, vitellogenin is phosphorylated during its time in the Golgi apparatus, followed by rapid excretion from the hepatocyte.

A similar situation appears to exist in the chicken, where three different genes of the vitellogenin family are expressed, producing three polypeptide chains with molecular weights ranging from 170,000 to 190,000. Moreover, the three subunits possess different degrees of phosphorylation and subsequently make up the native vitellogenin dimer. By analogy, a similar multigene family can be expected to code for vitellogenin in the fishes. The observed multitude of differing egg phosphoproteins further suggests widespread heterogeneity within the vitellogenin molecule.

TABLE 3.5

Selected Estradiol-Dependent Genes in Lower Vertebrates[a]

Proteins	Organ	Organism(s)	Effect
Ovalbumin	Oviduct	Birds	Induced
Lysozyme			Induced
Conalbumin			Induced
Ovomucoid			Induced
Avidin			Induced
Vitellogenin	Liver	Oviparous vertebrates	Induced
Albumin		*Xenopus, Oncorhynchus nerka* (?)	Depressed
ApoB, ApoII (VLDL)		Chicken, teleosts (?)	Induced
Vitamin-binding proteins		Birds	
Biotin			Induced
Thiamin			Induced
Cobalamin			Induced
Riboflavin			Induced
Transferrin			Induced
Estrogen receptor		Teleosts, chicken	Induced

[a]For structural and further biochemical changes initiated in the hepatocyte under the influence of estradiol.

Other Actions of Estradiol

The specific action of estradiol at the nuclear level in hepatic tissue is by no means restricted to the activation of the vitellogenin gene, although, at least in the fishes, vitellogenin constitutes the single most important de novo synthesis of protein. In the chicken, which has attracted most attention in this respect, a number of other genes are activated concomitantly, including those coding for apoVLDLII—the major VLDL (very-low-density lipoproteins) in laying chickens (*Deeley, et. al.,* 1985) — as well as various vitamin-binding proteins. In a mechanically unknown fashion, estradiol also induces the multitude of ultrastructural changes occurring in the liver of all vertebrates actively undergoing vitellogenesis. As pointed out above, the teleost fishes add another estrogen-induced gene to this growing list, namely, the gene for the nuclear estrogen receptor protein.

Estradiol also exerts negative effects on the synthesis of other export proteins, among them the ubiquitous serum albumins. In many vertebrates, estradiol administration leads to a pronounced reduction in the concentration of albumin circulating in blood, an effect that is especially apparent in chronically estradiol-exposed male *Xenopus*. Experiments conducted by Tata and co-worker led to the conclusion that in this amphibian, two levels of estrogen action on albumin synthesis can clearly be distinguished. First, estradiol administration leads to a deinduction of transcription of the two genes coding for the larger (74 kDa) and more abundant albumin. Second, it also causes a substantial destabilization of the messenger RNA for albumin, which is reflected in a decrease in the actual half-life of the messenger RNA by two-thirds. The same deinduction of albumin synthesis in the presence of estradiol can also be observed *in vitro* using isolated hepatocytes.

In fishes, a similar reduction in the amount of circulating albumin is apparent in naturally vitellogenic sockeye salmon (*O. nerka*) during their spawning migration. Conversely, in the rooster, estradiol *withdrawal* results in the destabilization of vitellogenin and apoVLDLII mRNAs, while the stability of the serum albumin mRNA is not affected. In addition to inducing the de novo synthesis of vitellogenin mRNA, estradiol has been shown to accelerate the rate of transcription of other genes, while not necessarily altering the *amounts* of mRNAs coding for different genes.

In all lower vertebrates, estradiol exerts a pronounced lipogenic action on peripheral tissues, while in *Xenopus* it also enhances the activities of enzymes involved in the hepatic synthesis of lipids. It can be speculated that the de novo synthesized lipid will be partly destined for the lipidation of vitellogenin and partly for inclusion in the increased output of VLDL by the liver. To date only one study has addressed this topic in fishes: female capelin (*Mallotus villosus*) displayed considerably higher total activities of fatty acid — *catabolizing enzymes* than did their male counter-parts. However, as long as only one part of fatty and metabolism (either anabolic or catabolic direction) is analyzed, no conclusive statement can be made about the lipid turnover in the respective tissues. Ultimately, the ratio of fluxes in the two directions will influence the actual net flux, and hence determine net import or export. The increased potential in vitellogenic grouper to generate cytosolic NADPH furnishes circumstantial evidence supporting the notion of increased hepatic fat synthesis during this period. In light of the general observations of increased lipid content in the blood of vitellogenic fishes, it can be speculated that while lipid turnover is stepped up, net flux is increased in the direction of lipid export from hepatic tissue. Strong lipogenic action of estradiol has been reported for *S. gairdneri* and *Heteropneustes* fossilis. Micrographs of vitellogenic livers of Fundulus heteroclitus contain less lipid depositions than livers from male fish while the livers of two other teleosts (*Notemigonus crysoleucas* and *Brachydanio rerio*) increased the amounts of lipid under the influence of estradiol in a dose-dependent fashion.

In the blenny (*Zoarces viviparus*), lipid is accumulated in the liver before vitellogenesis is hormonally induced, and subsequently, the lipid is mobilized and can be found in the bloodstream during vitellogenesis. Also, estradiol treatment during the course of pregnancy — a nonvitellogenic period in the blenny — leads to a dose-dependent accumulation in vitellogenin and a concomitant increase of lipid in the blood.

In conclusicn, two different strategies can be envisaged with respect to lipid mobilization and estradiol action in different species of fish. The simpler situation exists in fishes that accumulate lipids within the liver, such as the cod or the blenny. Here estradiol is likely to first cause a mobilization of intrahepatic lipid stores and later increase the output of VLDL from the liver. In fishes that use extrahepatic sites for lipid deposition, such as salmorids or a sculpin estradiol first induces the mobilization of extrahepatic lipids, and perhaps subsequently paces their uptake into the liver leading to increased hepatic output of VLDLs.

The treatment of goldfish with salmon gonadotropin leads to an augmentation of plasma triglycerides and cholesterol in goldfish with undeveloped ovaries, a phenomenon that is most likely mediated by gonadotropin-dependent estradiol production by the ovary. In animals undergoing the final stages of ovarian development, the same treatment decreases plasma lipid concentration, which is possibly due to a gonadotropin enhanced lipid uptake into the ovary.

Varied results are reported for the changes in intracellular glycogen content following *estradiol* treatment, although the generally observed trend seems to support the notion that hepatic glycogen is decreased in vitellogenic females. However, variable results for the contents of hepatic glycogen can be expected, since of all storage materials they are the most likely to be dependent on the preexperimental state of the experimental organism with respect to variables such as diet, photoperiod, and temperature.

Vitellogenic females of the killifish *F. heteroclitus* or estrogen-injected males contained less glycogen in their livers than uninjected male fish. A similar picture can be found in many other teleost fishes [*H. fossilis, Z. viviparus, S. gairdneri, Anguilla anguilla*], where estradiol-primed vitellogenic fish generally contain less glycogen in their livers than do vehicle-injected controls. In the grouper, in contrast, induction of exogenous vitellogenesis leads to a marked increase in hepatic glycogen. Sockeye salmon (*O nerka*) build up maximum liver glycogen levels at the end of the spawning migration, when exogenous vitellogenesis is approaching completion, and the fish subsequently call upon liver glycogen to fuel the exhausting spawning process.

An integral part of estradiol action is the observed hypercalcemia in vitellogenic fish, which can largely be ascribed to the calcium-binding properties of phosphorylated, and hence highly charged, components of the native vitellogenin molecule. Furthermore, this hypercalcemia has been employed to confirm the vitellogenic state of experimental animals. Fish scales have been singled out as the suggested source of the bound calcium, while, for once, estradiol does not seem to be implicated in the uptake of environmental calcium, neither through the gills nor though the intestine. As in the case of carotenoid binding, the actual site for the attachment of calcium to the vitellogenin molecule has not been identified, although liver seems the most likely candidate. If in the future it can be confirmed that other metals, such as copper or cadmium, travel from their hepatic deposition site to the ovary bound to vitellogenin, it will be appreciated how easily heavy metals will be able to impair the fine-tuned ion balance of the growing oocyte.

It is interesting to note that cortisol, a steroid hormone, which is known to exert direct metabolic effects by way of enzyme induction and permissive effects on peptide hormones such as glucagon, also possesses a pronounced enhancing influence on the estrogen-induced synthesis of vitellogenin in the catfish *H. fossilis*. This situation is somewhat reminiscent of the estrone-dependent priming of vitellogenesis through estradiol. Glucocorticoid administration to cultured hepatocytes curtails the vitellogenic response to estradiol, while at the same time enhancing the production of albumin, a protein whose synthesis may be suppressed in the presence of estrogen. Furthermore, recent experiments also suggest an important role for thyroxine, which is tightly bound by isolated fish liver nuclei, as an accelerating factor in exogenous vitellogenesis in the guppy *(Poecilia reticulata)*. Evidently, a number of other hormones interact in an as yet undetermined manner with estradiol during exogenous vitellogenesis, these interactions should provide a multitude of challenging topics of study for endocrinologists and molecular biologists.

Male Fish

It is an interesting facet of the induction of vitellogenesis that the estrogenic response can also be elicited in males of oviparous vertebrates, including fish. It clearly indicates that the administration of estradiol can activate normally silent genes. The complete absence of products of these unexpressed genes has made male animals a prime model for the analysis of gene regulation and activation. Basically, the male liver can be "reprogrammed" to synthesize and export large amounts of vitellogenin and other proteins, a process that appears to occur without involving DNA replication. Since an appropriate deposition site is lacking in the male, the fate of the vitellogenin in the bloodstream differs: it builds up to rather high concentrations and eventually is taken up by the liver and degraded along with other blood proteins.

The actual process of vitellogenesis is accompanied by identical patterns of hepatocyte differentiation in both sexes, including the proliferation in Golgi vesicles, rough endoplasmatic reticulum, and RNA mentioned. In the male Atlantic Salmon *(S. salar)* the estrogenic response also includes an increase in the amount of assayable nuclear estrogen receptor to levels characteristic of induced female fish, which is probably due to de novo synthesis of the receptor protein. The identical situation in hepatocytes from male *Xenopus* has made it possible to unequivocally identify receptor synthesis as the rate-limiting step in vitellogenin gene transcription.

In male fish, vitellogenin synthesis cannot be stimulated by the administration of pituitary extracts, indicating two specific properties of the vitellogenic response in fishes : (1) with regard to exogenous vitellogenesis, the liver is not a direct target organ for pituitary hormones, and (2) in males, vitellogenesis is specifically dependent on estrogen administration, because of the inability of the gonad to produce estrogen.

Elasmobranch Fishes

In general, vitellogenesis and its hormonal control in the elasmobranch fishes have received much less attention than in teleost fishes. The few studies on elasmobranchs suggest that, even in species from temperate zones, vitellogenesis and oviposition appear to occur throughout the year, with a maximum during winter. As a consequence, (vitellogenin is detectable in dogfish (*Scyliorhinus canicula* and skate (*Raja erinacea)* blood throughout the year, albiet in a low concentration compared with vitellogenic teleosts. The biochemical properties of the elasmobranch vitellogenins and their relationship to vitellogenins from other vertebrates remain to be analyzed.

Injection of estradiol results in a much smaller vitellogenic response than in teleosts. While the synthesis of vitellogenin in the female dogfish is a slow process compared with teleosts, its uptake into the ovary is fine-tuned to the rate of its synthesis. This results in an unusually long half-life for vitellogenin (9 days) in dogfish plasma, and a similar result can be expected for other elasmobranchs that are vitellogenic throughout the year. The only other systems where such long half-lives for vitellogenin represent the rule rather than the exception are the estrogen-primed males of other vertebrates that possess no tissue that would recognize vitellogenin for uptake. In male, estrogen-injected *Xenopus*, for instance, vitellogenin is removed from the bloodstream at a rate of less than 1% per day — which resembles plasma protein turnover — compared to more than 12% per day in the vitellogenic female.

4

Development

Fish eggs and larvae provide a relatively untapped source of biological material increased by the recent improvements in techniques for rearing marine species. Apart from their intrinsic interest, experimentally based information on these early stages is required for further progress in the advancing fields of fish culture and fisheries research. General text-books on ichthyology such as those of *Lagler et al.* (1962), *Nikolsky* (1963), *Norman* (1963), and *Marshall* (1965) and on reproduction in fish by *Breder* and *Rosen* (1966) provide both general and some detailed information. Identifications of *Ehrenbaum* (1909), *D'Ancona et al.* (1931-1933), through the current series of plankton sheets issued by the International Council for the Exploration of the Sea, and with the help of the extensive bibliographies by *Dean* (1916) and *Mansueti* (1954).

Most species of fish pass through a larval stage before assuming the adult form at metamorphosis. Sometimes the newly hatched fish is called a "prolarva" (or alevin in salmonids) until the yolk is resorbed, and then a "postlarva" (or fry). The term "larva" is used here for all stage to metamorphosis in marine fish, although alevin and fry may be used when referring to salmonids or other freshwater groups.

The Parental Contribution

Apart from the more obvious genetical effects on differentiation, rate of development, body form, size and behaviour, the parents, and especially the female, have an important influence on the vitality of the offspring both on a species and individual level, in terms of (a) the condition for incubation, (b) fecundity and (c) egg size.

Conditions for Incubation

Differences of spawning season and time and of spawning sites and substrate mean that incubation can take place in a great variety of conditions which influence the early development and physiology of the offspring.

1. Eggs Single, with No Parental Care

(a) Buoyant, planktonic —most marine fish, e.g., gadids, clupeids, flatfish and deep-sea fish.

(b) Nonbuoyant, loose or attached to substrate — a common fresh-water characteristic, e.g., cyprinids, pike Esox, or in littoral species, e.g., blenny *Blennius*, bullheads *Cottus*, sand eels *Ammodytes*; also found in some marine species, e.g., herring *Clupea harengus*, capelin *Mallottus villosus*, catfish *Anarhichas*, and American flounder *Pseudopleuronectes americanus*. Tendrils for attachment: are found in many oviparous elasmobranchs, in the hagfish *Myxine*, smelt *Osmerus*, saury *Scomberesox*, and flying fish *Exocoetus*.

(c) Nonbuoyant, buried in sand or gravel — many salmonids, grunion *Leuresthes tenuis*, and lamprey *Petromyzon*; in peat or mud *Aphyosemion* and *Cynolebias* where the eggs undergo diapause during the dry season.

2. Eggs Single, Special Environments

The bitterling *Rhodeus amarus* lays eggs in the gills of the fresh-water mussel and the lumpsucker *Careproctus* under the carapace of the Kamchatka crab.

3. Eggs Single, with Parental Care

(a) No nest, but eggs protected — found in many littoral forms, e.g., the bullheads Cottidae, blennies Blenniidae and gobies Gobiidae.

(b) Nests, often with parental protection and ventilation— also found in littoral species, e.g., blenny *Ictalurus*, sticklebacks

Gasterosteus, and in other freshwater species such as sunfish Centrarcidae, bowfin *Amia*, lungfish *Protopterus* and *Lepidosiren*, and in the Cichlidae. Bubble nests giving good aeration are found in tropical or swamp species, e.g., Siamese fighting fish *Betta splendens*.

(c) Parents carrying eggs — sea horses *Hippocampus* and pipefish *Syngnathus* have brood pouches and the sheat fish *Platystacus* a specially modified area of "spongy" skin. Marine catfish Ariidae, cardinal fish *Apogonidae* and *Tilapia* are mouth brooders, and *Tachysurus* incubates the eggs intestinally.

(d) Ovovipiparity and viviparity (see Section IV, C) — elasmobranchs include picked dogfish *Acanthias*, smooth hound *Mustelus vulgaris*, electric ray *Torpedo*, stingray *Trygon*, and the nurse hound *Mustelus laevis*. Teleosts include redfish *Sebastes*, *Heterandria*, *Anableps*, *poecinds* such as *Xiphophorus* and half beaks *Hemirhampus*.

4. Eggs Massed

Angler fish *Lophius* and yellow perch Perca flavescens have massed but unprotected eggs; in the lumpsucker Cyclopterus and butterfish *Blennius pholis* the eggs are protected by the male.

Fecundity and Egg Size

In higher latitudes the spawning season is often short and the eggs are liberated over a brief period of perhaps hours (clupeids) or over periods of some days, probably at certain times of the day or night (flatfish and gadids). Where the seasons are less marked spawning may occur over a much longer period or be intermittent throughout the year, especially where the time between generations is only a matter of weeks or months. Fecundity may be considered as the number of eggs produced in one year by a female although this may be very difficult to determine where spawning is protracted.

Some examples of fecundity, egg size, and length at hatching are given in Table 4.1. In general, fecundity is high where the

TABLE 4.1

Fecundity (Eggs/Female/Year), Egg Diameter, and Length at Hatching

Species	Common name	Fecundity	Diameter (mm)	Length at hatching (mm)
Molva molva	Ling	$20\text{-}30 \times 10^6$	1.0-1.1	3.0-3.5
Gadus morhua	cod	$20\text{-}90 \times 10^5$	1.1-1.6	4.0
Melanogrammus aeglefinus	Haddock	$12 \times 10\text{-}30 \times 10^5$	1.2-1.7	4.0-5.0
Pleuronects plastessa	Pliace	$16 \times 10\text{-}35 \times 10^4$	1.7-2.2	6.0-7.0
Soleu solea	Sole	15×10^4	1.0-1.5	3.2-3.7
Scouber scombrus	Mackerel	$35\text{-}45 \times 10^4$	1.0-1.4	3.0-4.0
Clupea harengus	Herring	$50 \times 10^2\text{-}20 \times 10^4$	0.9-1.7	5.0-8.0
Clupconclla delicatula	Kilka	$10\text{-}60 \times 10^3$	1.0	1.3-1.8
Saluo salar	Salmon	$10^3\text{-}10^4$	5.0-6.0	15.0-25.0
Osmerus eperlanus	Smelt	$50 \times 10^2\text{-}50 \times 10^3$	0.9	4.0-6.0
Acipenser sturio	Sturgeon	$80 \times 10^4\text{-}24 \times 10^5$	?	9.0
Cyprinus carpio	Carp	$18\text{-}53 \times 10^4$	0.9-1.6	4.8-6.2
Acanthurus triostegus	Convict surgeon fish	40×10^{3}[a]	0.7	1.7
Oryzias latipes	Medaka	20-40[a]	1.0-1.3	4.5-5.0
Scyliorhinus caniculus	Spotted dogfish	2-20	65.0 (length)	100.0
Lebistes reticulatus[b]	Guppy	10-50[a]	—	6.0-10.0
Zoarces viviparus[b]	Blenny	20-300	—	35.0-40.0
Sebastes viviparus[b]	Redfish	$12\text{-}30 \times 10^3$	—	5.0-8.0
Mustelus mustelus	Smooth bound	10-30	—	250.0
Squalus ocanthias[b]	spur dogfish	2—7	24-32	240-310

[a] Number per spawning, which may be repeated often in one year.
[b] Viviparous or ovoviviparous.

eggs are liberated into open marine waters; it is lower in fresh-water species and where there is parental care. There is also a strong tendency for fecundity and egg size to be inversely related.

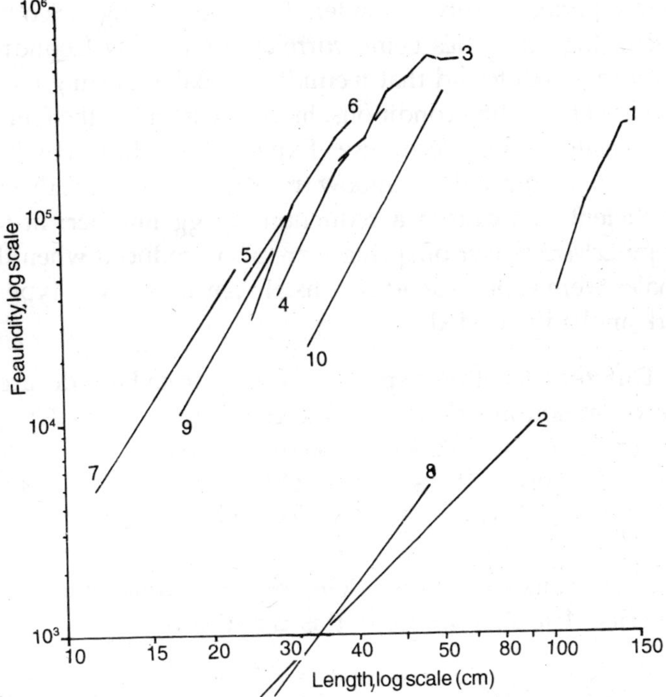

Fig. 4.1.   The relationship between fecundity and length within a species. 1. *Acipenser stellatus* ; 2. *Salmo salar*; 3. *Cyprinus carpio*; 4. *Pleuronectes platessa*; 5. *Clupea harengus* (northern North Sea); 6. *Melanogrammus aeglefinus*; 7. *Osmerus eperlanus*; 8. *Salvelinus fontinalis*; 9. *Sardinops caerulea*; and 10. *Sebastes marinus*.

Apart from enormous interspecific differences, there are also considerable variations of fecundity within a species. Many authors have found that fecundity increases with length, weight, or age the relationship usually being of the form F = aL^b, where F is fecundity, L is length, and *a* and *b* are constants. Year-to-Year differences resulting almost certainly from environmental effects are also well established. For instance, sea temperature was correlated by *Rounsefell* (1957) with the fecundity of pink salmon, *Oncorhynchus gorbuscha*, higher temperatures apparently resulting in lower fecundities. (Here the effect of temperature on growth is a complication). *Bagenal* (1966) reported density-dependent factors operating in Scottish flatfish, high densities being correlated with low fecundity. *Anokhina* (1960) found that fecundity in Baltic herring could be related to feeding conditions, high fat content of the female being related to high fecundity. Experimental studies by *D.P. Scott* (1961) indicated in rainbow trout, *Salmo gairdneri*, that an insufficient diet caused a reduction in egg number; in the guppy *Lebistes* fewer offspring were also produced when the females were kept on short rations. Extensions of this type of work are badly needed.

Differences within a species resulting from latitude, area, race or season are no doubt interconnected. Considerable differences of this type were reported for plaice, *Pleuronectes platessa*, for herring, for species of *Oncorhynchus* and for *Salmo salar*. An interesting characteristic of certain species is a difference in fecundity of the left and right ovary. The left gonad contains more eggs in *Oncorhynchus Salmo* and smelt *Osmerus*. The significance of this is not clear.

Egg size varies at the interspecific level (see Table 4.1), with larger eggs being especially associated with fresh-water species like to salmonids or where fecundity is very low, as in many elasmobranchs. As the intraspecific level, differences of egg size as a result of area or river were noted by *Rounsefell* (1957) in *Oncorhynchus, Salmo, Cristivomer*, and *Salvelinus* species. In *Salmo salar* there are differences in egg diameter related to length, fecundity, and river. In *Tilapia* egg weight may increase 2-4 times or even more depending on the size of the female in the flounder, *Platichthys flesus*, large females, or females from low salinities, have larger eggs. Larger eggs are also found in larger females of the spur dogfish *Acanthias*. There are differences in average dry weight of the order of four times among the various races of herring and small differences between very young first spawners and repeat spawners. In two darter *Etheostoma* species with long spawning seasons, the egg diameter tends to be greater in the cooler winter months.

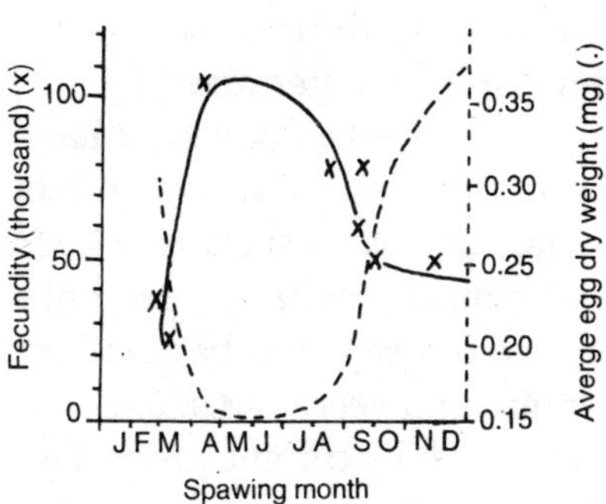

Fig. 4.2.   Fecundity and average egg dry weight in different races of herring.

The connection between egg size and fecundity in related species may be correlated with the conditions for incubation. For example, *Tilapia tholloni*, a substrate brooder, has 500-3000 eggs depending on length. On the other hand, *T. mossambica* and *T. macrocephala*, which are mouth brooders, have less than 500 eggs that are considerably greater in weight (*Peters*, 1963). *Garnaud* reported two species of *Apogon*, one, *A. imberbis*, with an egg diameter of 0.5 mm and fecundity of 22,000, and the other. *A. conspersus*, with an egg diameter of 4.5 mm and fecundity of 150. The relationship between egg size and egg weight *within* a species is shown for the different races of herring in Figure 4.2. The winter-spring spawners have a low fecundity and large eggs, an adaptation to poor food for the young, but a low predator population. In summer-autumn conditions fecundity is high and egg size low, presumably an adaptation to good food supplies and many predators. Intraspecific differences in fecundity and egg size deserve further study in terms of a link between ecological conditions and the physiology of maturation of the ovary.

A general pattern seems to emerge of marine fish with many, small buoyant eggs, a short incubation period, and vulnerable larvae. Fresh-water fish have larger demersal eggs, a long incubation period and larger, less vulnerable larvae, while the littoral forms exhibit protective devices to prevent losses in this particularly difficult environment. The initial conditions of development determined by the genotype of the parent and reproductive behaviour must have a considerable influence on the viability of the young and its physiology, in that environmental conditions such as temperature affect the speed of development, salinity presents problems of osmoregulation, and oxygen must be obtained for respiration. The egg is susceptible and yet cannot make defensive responses to mechanical shock, drift by current, toxins, light and predators. The larva should live in conditions where food can be obtained and protective behavioural devices can be practiced.

EVENTS IN DEVELOPMENT

A. Fertilization

The physiology of fertilization in fish, with special reference to the extensive japanese work, has been fully reviewed by *Yamamoto* (1961) from which much of the present account is taken.

There is some evidence for the action of gamones in *Lampetra* and in teleosts, these activate the sperm and serve as chemical attractants toward the egg, while other gamones are known to paralyze or agglutinate sperm. In the bitterling species *Acheilognathus* and *Rhodeus* sperm aggregation and activity have been noted in the micropyle region of the chorion.

The chorion or egg case is relatively tough with a funnel-shaped micropyle at the animal pole. Within the chorion a plasma or vitelline membrane [also called a pellicle or surface gel layer] surrounds the yolk and cytoplasm (ovoplasm) of the egg. Fertilization, which requires the presence of small concentrations of Ca or Mg ions, is normally monospermic in teleosts, the micropyle being too narrow to allow more than one sperm to pass at a time. The ovoplasm and chorion separate as the egg is activated by the sperm and a plug forms in the micropyle, further sperm being rejected. Where polyspermy occurs, as in some elasmobranchs, only one sperm fuses with the egg nucleus, the rest probably being resorbed and used at nutrient. Removal of the chorion permits polyspermy in teleost eggs; it seems that polyspermy is usually prevented by rapid changes at the micropyle, rather than over the egg cortex. In salmonids water activation (not to be confused with activation by sperm) takes place (e.g., see *Prescott*, 1955). When the egg is released into hypotonic solutions like river water the vitelline membrane becomes opaque and there are changes in its permeability. If sperm are not immediately available these changes may also affect fertilizability.

Following fertilization, the prominent alveoli in the egg cortex of salmonids, acipenserids, and lampreys disappear. In the medaka, *Oryzias latipes*, these alveoli break down progressively from the animal pole. The separation of the cortex from the chorion leads to the appearance of the perivitelline space. The chorion is permeable to water and small molecules, but larger molecules of a colloidal nature are retained in the perivitelline fluid. In *Oryzias* and *Lampetra* these colloids maintain an osmotically based tension within the chorion. It seems likely that the colloid is derived from polysaccharide material in the cortical alveoli so that the formation of the perivitelline space is partly owing to a decrease in volume of the ovoplasm, as the alveoli release colloid, and partly due to an osmotic distension of the chorion. According to *Ginsburg* (1961), polyspermy is blocked in sturgeon and trout eggs by the discharge of the cortical alveoli in the micropylar region. Chemicals, such as urethane, which cause polyspermy apparently retard the secretions of the alveoli, while removal of the perivitelline fluid in trout eggs allows the penetration of many sperm.

The chorion also hardens thus protecting the embryo in the early, more vulnerable stages. Probably the inner layer of glycoprotein is mainly responsible for this and it has been suggested that the alveolar colloid, Ca ions, phospholipids, or hardening enzymes also play a part. In salmonids, *Zotin* (1958) reported hardening of the chorion because of an enzyme in the perivitelline fluid; Ca ions affect the enzyme rather than the chorion itself. *Ohtsuka* (1960) considered that a phospholipid was liberated from the cortex (not from the alveoli) in *Oryzias eggs*. In *Fundulus* the chorion hardened with oxidizing agents but not with reducing agents. The soft chorion appeared to be impregnated with protein containing SH groups. Hardening resulted from oxidation of SH to SS groups by means of aldehydes produced from polysaccharides with x-glycol groups. *Zotin* distinguished between the initial enzyme action and subsequent hardening processes which last much longer and where the enzyme is no longer functionings. Thus, the initial enzyme reaction is blocked when Ca ions are bound by citrate or oxalate or by the use of NaCl or other chlorides. Later hardening is not susceptible to many of these factors.

The eggs of *Oryzias*, *Gasterosteus*, and *Lampetra* can be activated by pricking. Other activating agents are surface-active chemicals and lipid solvents (perhaps emulsifying the cytoplasm at the animal pole) and thermal shock, electric fields, ultraviolet light, and high frequency vibrations. This type of artificial parthenogenesis usually leads to irregular cleavage, but stringent precautions are needed to prevent contamination with sperm in such experiments.

Of interest is the ability of eggs and sperm to retain their fertilizability after leaving the parent. According to *Yamamoto* (1961) fish eggs lose this capacity after a very short time, but this can be increased if they are retained in isotonic Ringer's solution. While this is true of some freshwater eggs, presumably as a result of water activation, in seawater the capacity for fertilization is retained for much longer — certainly for hours in the herring. *Nikolsky* (1963) states that sperm motility is short lived where spawning takes place in fast-flowing water, for example, 10-15 sec in *Oncorhynchus*. In slower flows, sturgeon sperm is motile for 230-290 sec, and in the sea herring sperm may be motile for hours or days. Observations of short-lived activity are difficult to make; thus, some of these figures must be considered approximate.

Storage of gametes is a useful technique in fish farming to allow controlled fertilizations in the laboratory and to obviate the need for transporting eggs in the susceptible pregastrulation stages. Salmonid gametes are best stored dry below 5°C; those of the herring may be held in buffered egg-yolk diluents, but are also best kept dry at about 4°C. While this permits storage for, at most, a few days long-term techniques are also possible. Herring sperm, but not eggs, were kept for some months in a diluent consisting of 12.5% glycerol in 3% salt solution (diluted seawater) at — 79°C and crosses made successfully between spring and autumn spawning races. *Sneed* and *Clemens* (1956) also succeeded in holding out-of-season carp sperm immotile for 30 days at 3°-5°C in frog Ringer. Carp sperm could also be frozen and stored at — 73°C in isotonic Ringer containing 6-12% glycerol with some survival when thawed after 60 hss of storage. More recently, *Truscott et al.* (1968) have shown salmon that sperm can be stored for 1-2 months at temperatures of — 3°C to —4.5°C using diluents such as 5% ethylene glycol or 5% dimethyl sulfoxide, retaining 70-80% fertility. Horton et al. (1967) obtained alevins from salmon eggs fertilized with sperm frozen in liquid nitrogen with dimethyl sulfoxide as a protecting agent, but the fertility rate was low. *Hoyle* and *Idler* (1968) have also obtained fertile salmon sperm after storage in liquid nitrogen using ethylene glycol with added lactose or serine. Slow freezing produced the best results. *Mounib* et al. (1968) succeeded in storing cod sperm for up to 60 days using 17-24% glycerol and temperatures of —79° and —196°C. Initial experiments suggest that faster cooling rates gave the better results with cod sperm.

Incubation (Fertilization of Hatching)

The progress of cleavage, formation of layers, and morphogenesis have been described in a number of standard textbooks such as *Rudmick* (1955), *Waddington* (1956) and *Smith* (1957), with *Oppenheimer* (1947), and *Devillers* (1961) stressing structural changes from the viewpoint of experimental embryology. More detailed information is limited mainly to fresh-water species like the trout *Salmo trutta*, killifish *Fundulus*, medaka *Oryzias*, goldfish *Carassius*, and to the dogfish.

Most fish eggs are round, although in the anchovy *Engraulis* and *bitterling Rhodeus* they are avoid, and in certain gobies pear-shaped. Most species have telolecithal eggs with yolk more concentrated at the vegetative pole; some marine species have oil globules of varying size and number. Before fertilization the cytoplasm may be mixed with or separate from the yolk. The extent to which polarity exists at this stage has not been described in many species. After fertilization (as the cortical alveoli release colloid, the perivitelline space develops and the chorion hardens) cytoplasm migrates to the future blastodermal region, most of it arriving by the first cleavage. The remaining cytoplasm forms a "halo" or periblast under the blastoderm.

In lampreys cleavage is holoblastic but with the formation of micro-and macro-meres. In hagfish, elasmobranchs, and teleosts it is meroblastic. Other groups such as bowfin *Amia*, gar *Lepidosteus* and sturgeon *Acipenser* have intermediate features. With some variation, the meroblastic group possess a blastodermal cap of cells at the animal pole after the initial stages of cleavages. In *Fundulus* the surface gel layer overlying the blastoderm, being sticky on its inner surface, serves to hold the outer blastomeres together (*Triakaus*, 1951). Usually cleavage is not complete and in the deeper layers the periblast becomes syncytial and is involved in mobilizing the yolk reserves. There are substantial cohesive forces between the developing blastomeres and the surrounding periblast which are important in the subsequent morphogenetic movements.

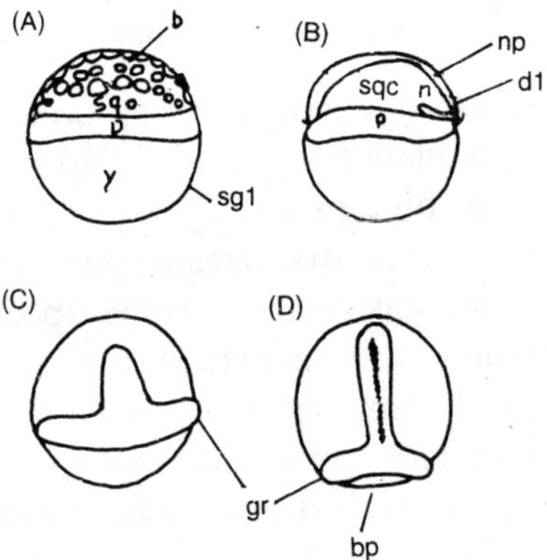

Fig. 4.3.   (A) Transverse section of early blastula showing adhesion of blastomeres to the surface gel layer and attachment of blastoderm to the periblast at the periphery only *(Fundulus,.* (B) Sagittal section of later blastula showing gastrulation; epiboly shown by arrows. (C) and (D) Surface view of eggs in later stages in gastrulation.   Key : b, blastoderm; bp, blastopore; dl, dorsal lip of blastopore; gr, germ ring; n, notochord; np, neural plate; p, periblast; sgc, sub-germinal cavity; sgl, surface gel layer; and y, yolk.

The blastoderm now commences to this and overgrow the yolk (epiboly) and at the same time invaginate at its periphery. The periblast seems closely connected with this spreading tendency of the blastoderm, which at the junction may be thickened to form a germ ring. The syncytial periblast seems to have the property of autonomous spreading, and it is likely that cell proliferation in the blastoderm is relatively unimportant. *Devillers* (1961) suggests that the periblast acts as an intermediary between two "non-wettable" components—the blastoderm and yolk. As epiboly proceeds the blastopore contracts as does the surface gel layer over the yolk. *In Fundulus* this layer probably solates and passes inward. The form of the developing germ then seems to be controlled by a balance of the forces of adhesion of the deeper blastomeres with each other and the syncytium, of tension in the surface (yolk) gel layer and of contractility of the·periblast.

The embryonic axis is laid down by a process of convergence and concentration in relation to the dorsal lip of the blastopore, the quantity of yolk present having some influence on the time at which this event occurs. Presumptive areas have been mapped in the early gastrulae of some species and show considerable variation. In earlier stages the eggs seem to be of the regulatory type. In *Fundulus* the 2-cell and 4-cell stage can withstand a loss of half the number of blastomeres. The embryos of *Carassius* can be divided at the 8-cell stage, each part sometimes giving a normal embryo. Up to the 16-cell stage two embryo can be fused resulting either in twinning or an oversized single embryo. Removal of the yolk from the blastoderm before a critical stage is reached (8 cell stage in *Carassius*, 32 cell stage in *Fundulus*, blastula in *Salmo*) brings development to a halt. Incomplete removal of the yolk before the critical stage may, however, not prevent further development. It is likely that organizer substances rather than nutrient material diffusing from the yolk are more important. At later stages even the embryonic shield of *Fundulus* may be isolated and cultured to a fairly advanced stage, with the development of primitive axial organs, ears and eyes, and even with cardiac contractions and independent movement (*Oppenheimer*, 1964).

Hatching

The time of hatching is both a specifically and environmentally controlled character with temperature and oxygen supply exerting a. considerable effect. Hatching results from a softening of the chorion because of enzymic or other chemical substances which are secreted from ectodermal glands usually on the anterior surface or from endodermal glands in the pharynx. In the sturgeon *Acipenser* the latter are innervated by the palatine nerve. The activity of the larvae, which may be enhanced by increase of temperature or light intensity or by reduction of oxygen tension, assists in breaking through the chorion.

The biochemical aspects of hatching are dealt with by *Hayes* (1949), *Smith* (1957), and *Deuchar* (1965). The enzymes have apparently been identified in a number of species, but certainly in salmonids Hayes' work shows there is doubt about their mode of action. The chorion, which resists digestion by trypsin and pepsin, appears to be of "pseudo-keratin." *Kaighn* (1964) measured the amino acid and carbohydrate components in *Fundulus*. Cystine comprised only 1%, compared with 12% in keratin. It is therefore unlikely that disulfide links play a role in stabilizing chorionic protein as they probably do keratin. The hatching enzyme works best under alkaline conditions. pH 7.2-9.6 and temperatures of 14°-20°C having been reported as optima. Very little hydrolysis takes place and Hayes speculates that the enzyme may be a reducing agent which liquefies the chorion. In *Oryzias* the enzyme is probably a *tryptase*. In *Fundulus*, Kaighn (1964) obtained purified chorionase and concluded that digestion of the chorion was mainly a proteolytic process, although he could not duplicate its action with other proteases. Whatever, the mode of action of the enzyme, it is likely that a considerable part of the nutrient material in the chorion can be utilized by the embryo via the perivitelline fluid and the losses at hatching may not be too serious.

The Larva

At hatching the larva is usually transparent, with some pigment spots of unknown function. Notochord and myotomes are clear with usually little development of cartilage or ossification in the skeleton. A full complement of fins is rarely present, but a primordial fin fold is well-developed in the sagittal plane. The mouth and jaws may not yet have appeared, and the gut is a straight tube. Although the heart functions for a considerable period before hatching the blood is colorless in the majority of species and the circulation and respiratory systems poorly developed. The yolk sac is relatively enormous with, presumably, hydrodynamic disadvantages. Pigmentation of the eyes is very variable, but where the eye is not functioning at hatching it very soon develops. The kidney is usually pronephric with very few glomeruli. Very little is known about the endocrine glands, gonads, and other organs of the body cavity at such an early stage.

As the yolk is resorbed, the mouth begins to function, the gut and the eyes develop further, and the larva becomes fitted for transfer to sources of external food. One of the earlier systems to develop is that responsible for locomotion and support, the primordial fin being fairly soon replaced by medium fins and the skeleton laid down. This is one of the better known aspects of later development because it is a system less easily damaged in such delicate organisms and because of the importance of meristic characters in racial studies of fish. Branchial replaces cutaneous respiration as the gill arches and filaments appear. The swim bladder may or may not be present during the larval phase. It is possible that this and the eyes, which are potentially dangerous in making the transparent larva visible, are silvered in such a way as to render them inconspicuous.

Metamorphosis

A clear change or metamorphosis from the larval to adult form is to be found in many species. In others there may be a number of less marked metamorphoses, e.g., in salmonids and eels. The most obvious signs are the laying down of scales and other pigmentation and often the first appearance of hemoglobin in the circulation. The swim bladder and lateral line may also develop first at this stage. In flatfish there is rotation of the optic region of the skull and the changes in the normal orientation of the body so that they eventually come to lie on one side. There are often concomitant changes in distribution and behaviour such as schooling. Barrington (1961) gives detailed consideration to the physiological changes associated with metamorphosis in salmonids, eels, and the lamprey, especially from the aspect of thyroid activity and osmoregulatory functions.

The time to reach metamorphosis may be a matter of days in tropical species, a few weeks or months in the majority of fish from temperate latitudes, or periods of years in the sturgeon *Acipenser* and eel *Anguilla*. It is controlled not only genetically but also by temperature and food supply, which may affect the rate of growth, and possibly by social (hierarchical) factors as well.

Timing

To give some idea of the timing of the events just described, examples of the approximate duration of different stages under natural conditions are given in Table 4.2. The modification of these times experimentally or by fluctuations in environmental conditions is discussed in the succeeding pages.

METABOLISM AND GROWTH

Rate of Development

Obvious specific differences in time to hatching may be masked by variations in ambient temperature, which is one of the most potent influences on rate of development. Detailed observations of temperature effects during development are scarce. *Fluchter* and *Rosenthal* (1965), however, showed between 3.5° and 9°C a doubling of heart rate in the embryos of the blue whiting, *Micromesistius poutassou*, and more rapid embryonic movements. The effect of temperature on time to hatching, a commonly used criterion, is shown in Figure 4.4. Low temperatures retard hatching) and at a theoretical low temperature, (the biological zero) the incubation period will be infinite. The product of incubation time (D) and temperature (T) — day-degrees — was originally thought to be constant i.e.

$$TD = k \tag{1}$$

This was modified to use the temperature, not from zero, but from the biological zero (T_o), i.e.,

$$(T - T_o)D = k \tag{2}$$

TABLE 4.2

Duration of Events in the Development of Some Species

	Weeks from fertilization to				
Species	Hatch	First feeding	Meta-morphosis	Temp. range (oC)	Main author
Sconer scombrus (mickerel)	0.8-1.5	1.3-2.0	11-13	9-15	Sette (1943)
Roccus saxatilis (straped bass)	0.25	0.75	4-5	17	Mansueti (1958)
Osmeral eperlanus (smelt)	2.0-4.0	2.5-5.0	8-10	4-14	Lillelund (1961)
Acanturus triostegus (convict surgeon fish)	0.15	0.7	?	26	Randall (1961)
Cluped harengus (Clyde herring)	2.5	3.5	16-18	7-10	Blaxter and Hempel (1963)
Pleuronectes platessa (plaite)	2.5	4.0	10-12	7-11	Ryland (1966)
Oryzias latipes (medaka)	1.5-2.0	2.0-2.4	Not clear cut	20-25	New (1966); T. Iwai, personal communication
Salmo salar (salmon)	20-22	26-28	Gradual	1-7	D.H.A. Marr, personal communication
Squalus santhias (spur dogfish)	ca. 104	ca. 104 (1944)	<104	4-12	Templeman
Scyliorhazus caniculus (spotted dogfish)	24-32	28-36	28-36	4-12	Amorosa (1960)

There has been increasing criticism of the concept, for example, by *Kinne* and *Kinne* (1962) and *Garside* (1966), on various grounds. Plots of $1/D$ against T are *curvilinear* over wide ranges of temperature, simple linearity only applying over a narrow range. This invalidates the formulas given above. Furthermore, there may be infections even of the curvilinear relationship at extreme temperatures. The biological zero, which is usually given between 0° and —2°C and most often around —1.5°C, may be below the freezing point of water or the body fluids themselves. In addition, abnormalities may occur at less extreme temperatures which are not necessarily lethal in the strict sense.

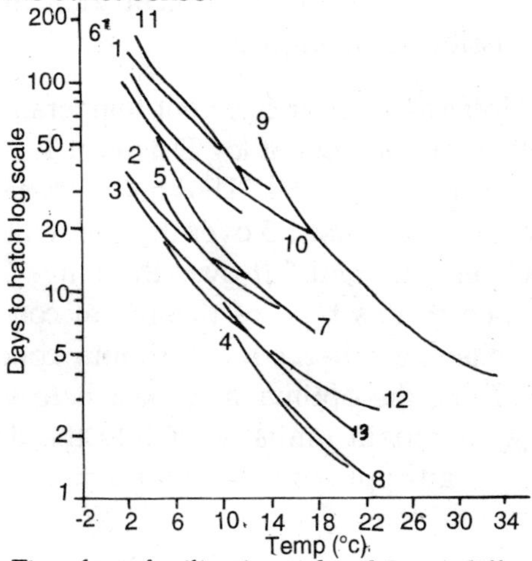

Fig. 4.4. Time from fertilization to hatching at different temperatures. 1. *Salvelinus fontinalis*; 2. *Pleuronectes platessa*; 3. *Gadus macrocephalus*; 4. *Sardinops caerulea*; 5. *Clupea harengus*; 6. *Coregonus clupeaformis*;; 7. *Osmerus eperlanus* ; 8. *Roccus saxatilis*; 9. *Cyprinodon macularius*;; 10. *Salmo gairdneri (irideus)*; 11. *Salmo trutta (fario)*; 12. *Enchelyopus cimbrius*; and 13. *Scomber scombrus*.

Improvements in describing the mathematical relation between D and T arise in later work. For example, *Blaxter* (1956) used the equation

$$(T - T_o) (D - D_o) = k \qquad (3)$$

for development of *Clupea harengus*, where D is the theoretical time to hatching at infinite temperature, not in itself a very satisfactory additional constant. *Lasker* (1964), working with the eggs of *Sardinops caerulea*, used the equation

$$D = aT^b \qquad (4)$$

where a and b are constants, are *Braum* (1964), using the eggs of whitefish *Coregonus* and pike *Esox lucius*,

$$D = Dx + 1.26^{Tx-T} \qquad (5)$$

where Dx is the minimum possible incubation time at the maximum permissible temperature T.

The van't Hoff values over different temperature ranges reflect the nonlinearity of plots of log D against T, the values being higher at lower temperatures. Thus in *Enchelyopus cimbrius* the Q_{10} varies between 6.5 and 1.5 over the temperature range 5°-23°C and between 6.5 and 2.0 over the range 3°-18°C in herring. The value of this type of theoretical consideration may lie in establishing criteria for optimum conditions of development. Thus, the optima may be where van't Hoff values lie between certain limits. Certainly the day-degree concept is useful as an approximation for predicting events in normal hatchery practice.

Other environmental factors influence the rate of development. Low salinities may accelerate or retard the time to hatching while oxygen lack has a retarding effect on development, especially at higher temperatures. *Laale* and *McCallion* (1968, found that the development of the zebra fish, *Brachydanio rerio*, could be arrested before gastrulation by the use of the supernatant of homogenates produced from other zebra fish embryos. This arrest, which could be reversed, appeared to be an effect at the cellular level, the nuclei of the arrested embryos all being in interphase.

Yolk Utilization

The efficiency with which yolk is transformed to body tissue and the effect of the environment on utilization is important in that larger larvae may be expected to be stronger, better swimmers, less susceptible to damage, and less liable to predation. Efficiency at any time may be expressed as a percentage :

$$\frac{\text{dry weight increment of body}}{\text{dry weight decrement of yolk}} \times 100$$

More often efficiency is measured from fertilization to final yolk resorption (or to maximum weight attained on the yolk reserves). This is gross efficiency, i.e.,

$$\frac{\text{dry weight of final body}}{\text{dry weight of original yolk}} \times 100$$

or from fertilization to intermediate stages as

$$\frac{\text{dry weight of body}}{\text{dry weight of original yolk} - \text{dry weight of remaining yolk}} \times 100$$

or more precisely

$$\frac{\text{dry weight of body}}{\text{dry weight of body} + \text{dry weight of yolk used for maintenance}} \times 100$$

The difficulties of measuring efficiency by dry weight lie in the need for taking samples of an egg population at different stages with the accompanying problems of initial differences in egg weight. Utilization of material from the chorion or losses of excretory products are also difficult to allow for, as are the possibilities of uptake of organic matter from the environment Another serious problem when comparing, for example, environmental effects such as temperature on efficiency is the question of making dry weight measurements at "equivalent" stages. Both hatching and maximum weight (attained on the yolk) can be questioned for staging, hatching at different temperatures can result in larvae of quite different appearance, while full yolk utilization is often not complete when maximum weight is reached, some yolk remaining in the yolk sac. Furthermore, the disappearance of the yolk sac is no certain indication that all the yolk has been used as it may be present in storage spaces within the larval body, for example, in the subdermal spaces of cod and plaice larvae. *D.H.A. Marr* (1966) adopted the ratio as a criterion for *equivalent staging,* comparing in *S. salar* the efficiency

$$\frac{\text{dry weight of body}}{\text{dry weight of body} + \text{remaining yolk}} \times 100 \text{ (as percentage)}$$

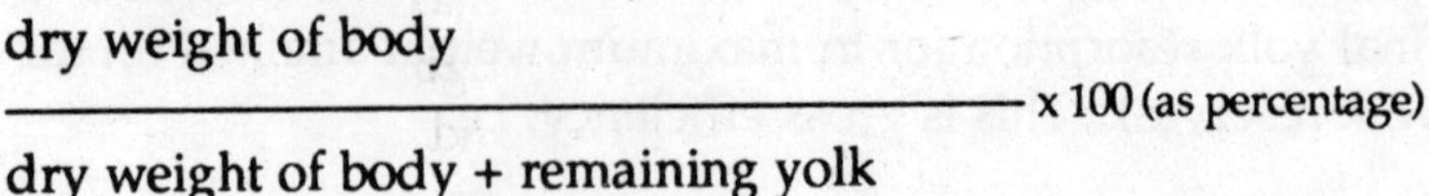
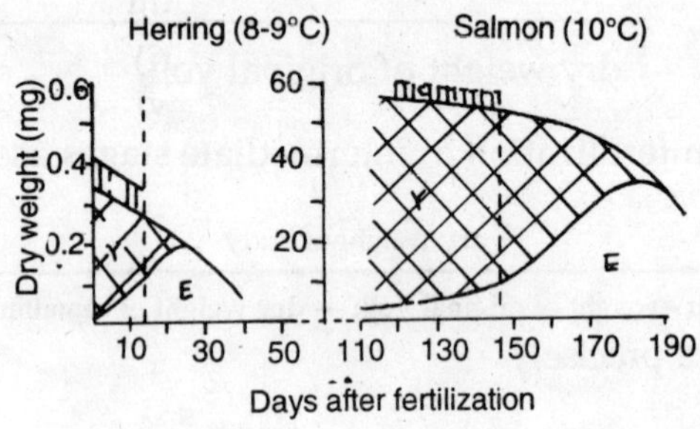

Fig. 4.5.   The relative proportions of yolk (Y), embryo (E), and chorion (C) during the development of a small egg (herring : Blaxter and Hempel, 1963) and a large egg. (salmon : D.H.A. Marr, 1966). The vertical dashed line represents hatching. Note the differences in the scale of the ordinates.

of development at different temperatures between the 15% and 80% stages. *Ryland* and *Nichols* (1967) used the ratio

$$\frac{\text{rate of growth in length}}{\text{rate of yolk disappearance}} \times 100$$

for equivalent staging when comparing the efficiency of development at different temperatures during the yolk sac stage of *P. platessa.* The use of maximum weight as a stage for making comparisons still remains, however, a useful criterion and one with immediate meaning when deciding on the optimum conditions for hatchery practice.

Calculations of efficiency by various methods are given in

Table 4.3. Efficiency over the whole process of yolk utilization is mainly between 40% and 70% although clearly cumulative efficiency must decrease as growth proceeds and the maintenance requirements increase. Experiments with temperature indicate certain optima for maximum efficiency. Other influences on efficiency are the original egg weight at the intraspecific level in *C. harengus*, and light conditions, contour of the substrate, and turnover of water in the photonegative alevin of *S. salar* living within the interstices of the spawning redd. Here highest efficiency is achieved under dark conditions, on a grooved substrate, with a rapid turnover of water.

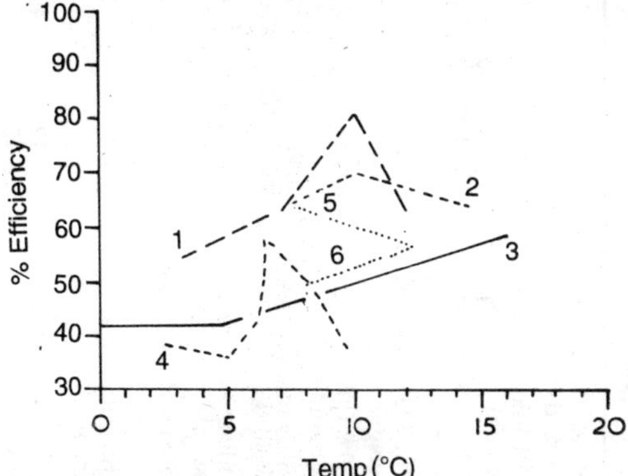

Fig. 4.6.   Efficiency of development (see text and Table III), at different temperatures. 1. *Salmo trutta (fario)* — yolk sac period (see D.H.A. Marr, 1966); 2. *Salmo salar*—alevin stage (D.H.A. Marr, 1966); 3. *S. salar*—early alevin (Hayes and Pelluet, 1945); 4. *Pleuronectes platessa*—yolk sac larva (Ryland and Nichols, 1967); 5. *Cluvea harengus*—small eggs, yolk sac period; 6. *Clupea harengus*—large eggs, yolk sac period (Blaxter and Hempel, 1966).

A word of caution is required where larval feeding may occur well before final yolk resorption. High efficiency may result from low activity, a high proportion of yolk being used for growth: if this is reflected in low feeding activity it could be a very undesirably trait.

TABLE 4.3
Efficiency of Yolk Utilization

Species	Method	Stage	Temp. (oC)	Efficiency (%)	Author
Salmo trutta (trout)	Dry weights	Fertilization to 50-80 days	10	63	Gray (1926)
Salmo trutta (trout)	Dry weights	Fertilization to max. weight	15	56	Gray (1928a)
Salmo salar (salmon)	Dry weights	Hatching to 10 days after	0-16	42-59	Hayes and sw Pelluet (1945) (Fig. 4.6)
Salmo salar (salmon)	Calorific values	Fertilization to max. weight	10	41	Hayes (1949)
Salmo salar (salmon)	Special dry weight % index (see text)	15-80% (yolk sac stage)	7.6-14.3	64-70	D.H.A. Marr (1966) (Figure 4.6)
Salmo gairdneri (rainbow trout)	Dry weight and metabolic criteria	Fertilization to max. weight	10	60	Smith (1957)
Fundulus heteroclitus (killifish)	Wet weights	Fertilization to max. weight	19.4-21.4	62	C.G. Scott and Kellicott (1916)
Silurus glanis (sheat fish)	(?)Wet weights	During yolk sac stage	?	66	Ivlev, quoted by Lakser (1962)
Sardinops caerulea (California sardine)	Calorific values and respiration	Fertilization to yolk resorption	14	79	Lasker (1962)
Clurpea harengus (herring)	Dry weights	Fertilization to hatch / Fertilization to max. weight	8-12 / 8-12	40-80 / 50-65	Blaxter and Hempel (1966) (Figure 4.6)
Pleuronectes platessa (plaice)	Rate of growth in length rate of yolk disappearance	During yolk sac stage	2.6-9.8	35-58	Ryland and Nichols (1967) (Figure 4.6)

Viviparity

Amoroso (1960), who gives a comprehensive review of this subject, points out the rather indistinct barrier between ovoviviparity where the young develop within the female only on their yolk reserves, and viviparity, where the nutrient requirements are obtained from the mother. In the first place, there are a wide variety of methods of obtaining these nutrients, by absorption through simple external surfaces, by swallowing, or by "placental" connections. Perhaps all these may be considered as viviparous. In the second place, initial development may be ovoviviparous (on the yolk supply) with viviparity superimposed later. This is likely to be the case in the very early stages of *all* viviparous fish, but in the smooth dogfish, *Mustelus laevis,* and goodeid teleosts, for example, there is a change from one to the other rather later. Details of the functional morphology of viviparity are dealt within the chapter by *Hoar*, this volume.

The changes of weight found during the development of various species can be very striking. Some of the ovoviviparous ones, where maternal nutrients are scarce, show decreases of organic matter between the fertilized egg and final embryo, for example, of 23-34% of *Torpedo* spp. This gives an efficiency of 66-74% which puts this species very much in the same category as oviparous forms although it is rarely certain what nutrient is obtained from the mother. The long gestation period of 4-6 months with this order of efficiency suggests some nutrients are being absorbed from the oviduct. In other "ovoviviparous" species there may be gains in organic matter, e.g., of 356% in *Mustelus vulgaris* and 1628% in *Trygon. In Mustelus laevis,* sometimes more strictly called viviparous, the gain is 1064%, although *Te Winkel* (1963) reported a fall in weight of organic matter early in the development of a close relative, *M. canis.*

Biochemical Aspects

Much work has been done in the past on the larger salmonid eggs which can provide a greater bulk for analysis or on whole ovaries. Some more recent work which include the use of isotopes, chromatography, and histochemistry is mentioned by *Deuchar* (1965) and *Williams* (1967).

Water Relations

The swelling of eggs, with the formation of the perivitelline space as a result of water uptake, is a general phenomenon signifying fertilization. The chorion at fertilization is permeable to water and also to urea, glucose, salts, and certain dyes. It seems likely that the colloidal material liberated from the cortical alveoli cannot escape through the chorion and creates an osmotic pressure which draws in water. This effect can be inhibited by high osmotic pressure in the outside medium. The chorion then hardens, a process taking a matter of a few hours and the osmotic forces become matched by the resistance of the chorion. In hypertonic solutions the eggs of salmonids and of *Mullus barbatus* lose water only from the perivitelline space and not from the embryo. Some loss does, however, occur in acipenserids. The use of D_2O on water-activated and early fertilized eggs of *Oncorhynchus tshawytscha* seems to confirm the view that only the perivitelline space is penetrable by water. Subsequent use of ^3H_2O, $^{22}NaCl$, and $Na\ ^{31}I$ on the water-activated eggs of *Salmo gairdneri* has shown a definite but limited permeability of the vitelline membrane to anions, cations, and water. Unfortunately, this was not done on fertilized eggs. Recently, however, *Potts* and *Rudy* (1969) have confirmed with ^3H_2O that the vitelline membrane of fertilized eggs of *S. salar* has a high permeability before laying and during water hardening. Subsequently permeability is low until the eyed stage. The use of ^{24}Na showed that sodium exchange is confined initially to the perivitelline fluid but accumulation within the embryo occurs during the eyed stage. *Terner* (1968) reported that during the eyed stage of *Salmo gairdneri* external substrates such as ^{14}C-labeled pyruvate and acetate were apparently taken up and metabolized as judged by the presence of $^{14}CO_2$ in the respiratory CO_2. *Wedemeyer* (1968) also found that ^{65}Zn was taken up by developing eggs of coho salmon *O. kisutch*. Almost all was bound to the chorion, but 26% was found in the perivitelline fluid, 2% in the yolk, and 4% in the embryo. *Mounib* and *Eisan* (1969) found that both ^{14}C-labeled pyruvate and glyoxylate could be utilized in the form of an exogenous substrate by salmon *(S. salar)* eggs. Lactate was produced and any carbon atom in these compounds could be incorporated by the eggs into organic acids, lipids, nucleic acids or proteins. The formation of ^{14}C-amino acids indicated the presence of an active transaminase system.

The use of freezing point measurements on the yolk of herring and plaice eggs gives an alternative picture. In the herring which has a demersal egg, the yolk is not regulated osmotically until after gastrulation when it is covered by cells. However, in the pelagic floating egg of the plate, regulation occurs from fertilization, indicating the ability of the vitelline membrane to osmoregulate. This seems necessary from the buoyancy aspect.

If Gray's (1926) measurements of wet and dry weights of the embryo and yolk of trout are generally true of salmonids, then the yolk as a high density nutrient (41% dry weight) requires considerable quantities of water for transformation to the relatively watery embryo (16% dry weight). Obtaining this water may be a problem if the permeability of the vitelline membrane remains limited throughout much of development. *Smith* (1957) suggests that growth may be retarded until hatching occurs and water becomes more readily available. It is likely that highly desiccated yolk is only feasible in the demersal egg; in *Sardinops caerulea* the water content of the larval yolk is about 91% (Lasker, 1962).

Chemical Composition

The chemical composition of the eggs of two species are given in Table 4.4. Further data are given by *Phillips* and *Dumas* (1959), who showed in particular that there was no difference in the constituents of different sized eggs of *Salmo trutta*.

TABLE 4.4

Analyses of Eggs

Species	Dry weight %	Protein	Total fat	oil	Phospholipid	Carbohydrate	Ash
			Percent of wet weight				
S. gairdneri (irideus) (rainbow trout)	33.8	20.2	—	3.6	3.8	0.2	1.3
Sardinops caerulea[a] (California sardine)	29.3	21.0	3.8	—	3.2	<0.3	2.1

[a] Data for trout from Smith (1957) and for sardne ovary from Lasker (1962).
[b] Calorific value of dry sardine-yield was 5.4 cal mg.

There is considerable difficulty in deciding on the sequence of utilization of various materials for energy production. Analysis of the change in chemical components may be unreliable where substances like carbo-hydrates are being synthesized. Heat production and respiratory quotients are difficult to determine accurately in small eggs, and CO_2 liberation may be masked by a buffered external medium. Amberson and *Armstrong* (1933) found the RQ's of *Fundulus* eggs over the first 6 days of development were 0.90, 0.78, 0.77, 0.76, 0.72 and 0.72 suggesting very early carbohydrate metabolism. In *Oryzias, Hishida* and *Nakano* (1954) found 0.75 after fertilization, 0.92 at gastrulation, and 0.70 later. Another problem to overcome is the retention of nitrogenous excretory matter within the egg which gives values of protein metabolism which are too low. The original concept of *Needham* (1931) that materials were used up for energy in the sequence

carbohydrate—protein—fat

has to some extent been supported by Devillers (1965). However, *Hayes* (1949) was of the opinion that the sequence in salmon eggs was

hatch
|

fat—protein—fat—protein

and Smith (1957, 1958) that it was

hatch

phospholipid—protein + carbohydrate—protein—phospholipid—triglyceride fat

Further work is required in this field, especially of a comparative nature.

Carbohydrate Metabolism

Since carbohydrate is present in small quantities and synthesis is continuously occurring, its role is particularly difficult to ascertain by bulk analysis. *Hayes* (1949) reported a gradual increase in carbohydrate level during salmon development with a temporary fall in glucose level at hatching. Glycogen is probably synthesized near hatching and is present exclusively in the embryo, being stored in the liver near the end of the yolk sac stage. Glucose, in aqueous phase, is present in relatively greater quantities in the embryo than in the yolk due to its higher water content. *Smith* (1957) described falls in total carbohydrate after gastrulation (linked with the establishment of the circulatory system and therefore a higher potential for metabolism) and a further fall in glucose level at hatching, perhaps because of an interruption of glycogen synthesis. He did not report the general increase found by Hayes. *Terner* (1968) has more recently found that the ova of *Salmo gairdneri* have a free store of glucose which increases during development and only falls at hatching. Carbohydrate is certainly being synthesized in development for use in the final stage of energy production and for mechanical and osmoregulatory work.

Protein

Protein is the dominant raw material in the yolk and the main source for tissue formation. The use of labeled amino acids in *Oryzias* shows a slow passage from the yolk to a low molecular weight pool, a more rapid uptake occurring when the vitelline circulation is established. The proportion of yolk used for energy requirements is not easy to determine since the egg retains some nitrogenous metabolities (shown by an increase of non-protein nitrogen with age) and only small amounts of ammonia are excreted. Of particular interest is the finding by *Read* (1968) of ornithine carbomoyltransferase and arginase in the embryos of *Squalus suckleyi* and *Raja binoculata*, suggesting an early functioning ornithine urea cycle with the retention of urea for osmoregulatory purposes. The use of protein for energy seems to decline after hatching, while the amount used over the whole of early development depends on whether calculations are made when some yolk remains or whether, at a later stage, the larva starts to consume its own tissues. While it has been suggested that 40% of the original protein may be metabolized more recent detailed studies using isotopes on Oryzias indicate that all protein being resorbed from the yolk in the 9 days subsequent to gastrulation was being transformed into new tissue and none lost in combustion.

Regarding different types of protein it appears that in *Oryzias* 40% of phosphorus is incorporated as phospho-protein: it falls after gastrulation and the loss is very rapid at hatching, especially in *S. gairdneri (irideus)*. *Deuchar* (1965), reviewing analyses of amino acids present in the same species, reports that aspartic and glutamic acid are always present, valine and leucine appearing after gastrulation.

Fat

There is some ambiguity about the role of fat in energy production during different phases of development. Fat is undoubtedly used as a fuel, possibly 70-90% being consumed over the whole period of development. *Hayes* (1949) reported the main loss of fat to be in the fourth week after hatching in the salmon. It seems that the triglyceride fats are the last to be utilized-before food is required from external sources. The energy requirements before hatching may also be met by fat, according to Hayes, or by fats in combination with protein.

According to *Phillips* and *Dumas* (1959) fat is synthesized from protein in the eyed stage of the eggs of brown trout *Salmo trutta. Terner et al.* (1968) found that C-labeled acetate in the incubation medium of the eggs of *S. gairdneri, S. trutta,* and *Salvelinus fontinalis* was incorporated into free fatty acids and other lipids of the egg, which were probably the substrates for endogenous respiration. In addition, labeled acetate was found in the egg phospholipids, presumably as an intermediate stage in the synthesis of complex fats. These authors suggest that the lipids of the embryo are not transferred directly from the yolk but are resynthesized by the embryo after the breakdown of triglyceride fats in the lipid pool of the yolk.

Respiration

In the sense of obtaining oxygen, respiratory problems vary with the environment of the egg and larva, but they are probably rather rarely limiting with pelagic eggs. Eggs deposited on or in a substrate may act as an oxygen "sink" the oxygen level at the egg surface always being less than that of the surroundings even in high velocities of current. Hayes *et al.* (1951) calculated a flow of $1.2 \times 10 \ \mu l/cm/cm^2$ thickness/min through the chorion of salmon eggs. Sensitivity to low oxygen tensions or anoxia varies with both species and stage of development. In salmonids, cleavage may continue in anoxic conditions, perhaps this is possible because requirements are low and oxygen is stored within the egg, especially in the perivitelline fluid. Gastrulation is more easily blocked by lack of oxygen or by respiratory poisons such as cyanide or azide. Development can also be retarded by blocking oxidative phosphorylation, with dinitrophenol. During oxygen lack there are often increases in lactic acid production but, in the short-term, retardation of development to anoxia is reversible. *De Ciechomski* (1965) succeeded in keeping the demersal eggs of *Austroatherina* for 17-18 days in "Vaseline oil." Development was at first normal, but no heart beat or movement was observed. Transfer to water resulted in the heart starting to beat and some movements occurred, but abnormal pigmentation developed, and no embryos hatched. Eggs of three pelagic species were killed by immersion in the oil.

Various adaptations are found which assist in obtaining oxygen. The spawning redd of the salmonid fish has a rapid current of water passing through it; parents guarding their eggs may ventilate them (e.g., the stickleback, *Gasterosteus aculeatus*) while the lungfish *Lepidosiren* has special pelvic "gills" for oxygenating the eggs in its nest; viviparous species obtain oxygen as well as nutrients from the female. After hatching functional gill filaments are often lacking; in herring they do not appear for some weeks at a length of about 20 mm. Oxygen is obtained over the body surface and almost certainly via internal body surfaces such as the pharynx and gut. The ultramicroscopic corrugations described by *Jones et al.* (1966) and Lasker and *Threadgold* (1968) on the epidermis might possibly be directing the flow of water over the body surface. In less favourable environments external gills may be found, for example, in the bichir *Polypterus*, in the lungfish *Protopterus* and *Lepidosiren*, in the loach *Misgurnus fossilis*, and in *Gymnarchus niloticus*. In newly hatched trout, *Salmo trutta*, respiratory currents are produced by the pectoral fins which also keep the body clear of silt. After a short time mouth and gill movements replace those of the fins and then clogging of the gills by silt is prevented by a coughing reflex or aggregating the silt with mucus.

Intake of oxygen has been measured in a number of species. In the egg stage this may be very low and variable before fertilization. In *Oryzias latipes* there is no sharp rise until a few hours after fertilization, although it seems likely that the respiratory rate becomes steadier at this stage (*Nakano*, 1953). Before any movements occur within the egg, relative values of oxygen uptake between different stages and species are quite valid, but subsequent to movement being possible the oxygen consumed can be dominated by bursts of activity. Oxygen uptake may be used to express metabolic functions, total or general metabolism consisting of two components, active metabolism (owing to activity) and basal metabolism (owing to maintenance functions). Standard metabolism applies to practically motionless fish under experimental conditions; in the fry of *Salmo salar*, for example, the basal metabolism was 64-69% the standard rate. For active fish metabolism is proportional to KV^n, where V is velocity and K and n are constants. The change in oxygen uptake with varying velocity is shown in Figure 4.7A. The function KV^n is called the "scope for activity" and has been estimated for a number of species (see Table 4.4).

Some values of oxygen uptake expressed as Qo_2 ($\mu l/mg$ dry weight/hr) are given in Table, 4.5 based where possible on subjective criteria of activity and inactivity. Without the type of precise control as demonstrated in Figure 4.7A these are the best available data at the present time. The standard Qo_2's show some similarity considering the variety of the material used. Active Qo_2's will naturally fluctuate widely, depending on the type and extent of activity permitted. The use of certain anesthetics to control activity may also help in establishing at least some of the factors controlling oxygen uptake.

Cumulative oxygen uptake has been measured over short periods as well as fairly long periods of development. In *Fundulus heteroclitus*, the killifish, it was 80 $\mu l/egg$ from fertilization to hatching (Scott and Kellicott, 1916). In salmon eggs (Hayes *et al.*, 1951) the uptake was about 0.2 ml/egg/hr at fertilization rising to 3.4 ml/egg/hr at hatching, the cumulative total between these stages amounting to 1400 ml/egg. Privolniev, quoted by Devillers (1965), found only 850 ml/egg from fertilization to an age of 120 days. Such values may be compared with the short yolk sac period of the California sardine, *Sardinops caerulea*, where only 12 ml/larva were used from hatching to 180 hr subsequently Lasker and Theilaekar, 1962). Obviously cumulative uptake and uptake per unit time will vary with size, age, and temperature. *Devillers* (1965) considers that

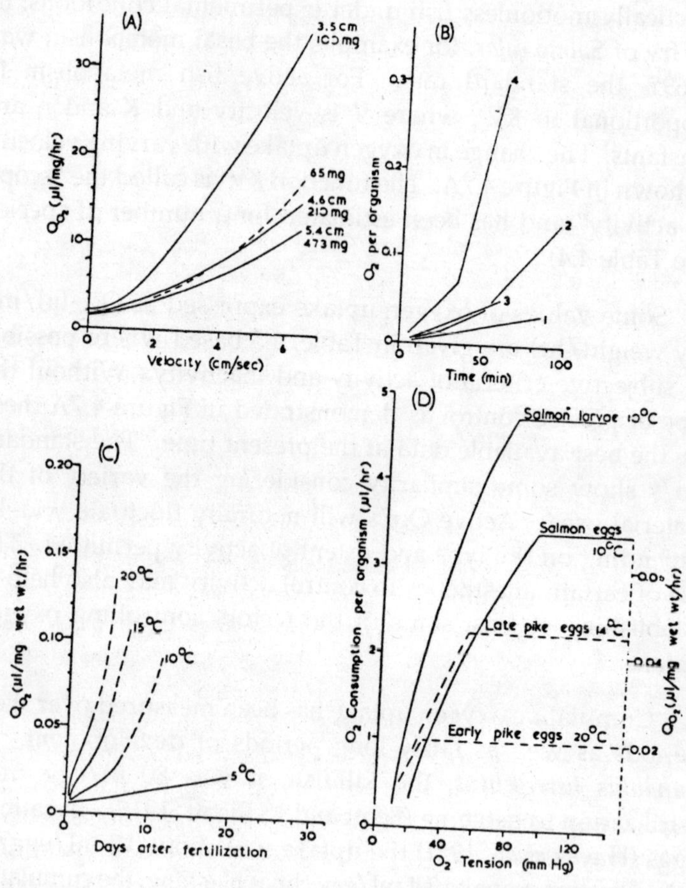

Fig. 4.7.   (A) Oxygen uptake (Q) for various sizes of young fish stemming different velocities of current at 20°C. (——) *S. salar*; (- - -) *Alburnus alburnus*. Wet weights of original data estimated as four times the dry weight. Estimated dry weight and the length given at the end of the each curve.   (B) Cumulative oxygen consumption in clupeid eggs and larvae. *Sardinops caerulea*: 1. 20 hr embryo at 14°C, 2. newly hatched active larvae at 14°C; *Clupea harengus*: 3. newly hatched inactive

larva at 8°C, 4. newly hatched active larva at 8°. (C) The effect of temperature on oxygen consumption (Qo_2 based on *wet* weight) of different stages of the eggs of *Esox lucius;* hatching at about 11 days at 10°C. (D) The effect of oxygen tension on oxygen consumption. (—) and left-hand ordinate, eggs and larvae of salmon, *S. salar;* (- - -) and right-hand ordinate, eggs of pike, *Esox lucius.*

TABLE 4.5

Scope for Activity

Species	*Stage*	*Range of increase of Qo_2 owing to activity*	*Author*
Salmo salar (salmon)	Eggs	3	Hayes *et al.,* (1951)
Salmo salar (salmon)	Fry	7-14	Ivlev (1960a)
Sardinops caerlea (California sardine)	Larvae	3	Lasker and Theilacker (1962)
Clupea harengua (herring)	Larvae	9.10	Holliday et al. (1964)

increases in oxygen uptake with age may be shown by the exponential equation

$$Q = ae^{kt} \tag{6}$$

where *t* is time and a and k are constants. Qo_2's will also vary with environmental conditions (see Figure 4.7C). In herring eggs and larvae the Q_{10} for oxygen uptake (Q_{02}) was 2.0 between 55° and 14°C. Q_{02} also increased with a salinity shock but not with constant rearing salinities. There were only slight effects because of light intensity and starvation. In salmonids oxygen uptake falls at low ambient oxygen tensions, but the young stages of fish will resist oxygen lack quite well for a time.

This raises the problem of oxygen debt in such experiments and indeed in any respiration measurements where activity is intense. The accumulation of such a debt can be tested by continuing respiration measurements after activity or anoxia.

TABLE 4.6

Oxygen Uptake as Q_{O_2} (lμ/mg dry wt/hr)

Species	Stage	Temp. (°C)	standard rate	Rate not defined	General (active) rate	Author
Salmo trutta (brown trout)	Egg, embryo only	10	0.55			Gray (1926)
S. salar (salmon)	Egg, embryo only					
	19 days	10	0.9			
	50 days	10		1.0		Hayes et al. (1951)
S. salar (salmon)	Fry (no yolk)					
	3.5 mm	20	2.3		35.0	Ivlev (1960a)
	4.6 mm	20	2.0		15.3	
	5.4 mm	20	1.6		12.1	
Fundulus heteroclitus (killifish)	Egg, embryo only	19-21	3.2			C.G. Scott and Kellicott (1916)
	6 day larvae (no yolk)	19-21		3.9		
Esox lucius (pike)	Whole eggs (range from) fertilization to hatch	5	0.02-?			Lindroth (1942)
		10	0.02-0.36			
		15	0.02-0.41			
		20	0.02-0.54			
Sardinops caerulea	Whole eggs	14	0.8		1.79	Lasker and Theilacker (1962)
(California sardine)	Whole larvae	14	1.33		2.68	
Clupea harengus (herring)	Whole eggs	8	1.5			Holliday et al. (1961)
		12	4.0			
	Larvae (less yolk)	5	2.0		2.5	
		14	3.5		5.0	

[a] Dry weight estimated at 16% of wet weight.
[b] Dry weight estimated at 25% of wet weight.

Hayes et al. (1951) found no oxygen debt after periods at reduced oxygen tensions in salmon eggs, suggesting there had been a reduction in metabolism under these conditions. Salmon fry however, an oxygen debt amounting to 48% of general metabolism after a period of activity. Fry of bleak, *Alburnus alburnus*, had a negligible debt when stemming a water velocity of 2 cm/sec but one of about 15% in 6.42 cm/sec.

An overall review of respiration rates, including that of larval fish, allowed *Winberg* (1960) to deduce that the relationship

$$Q_s = 0.3 \qquad w^{0.8} \tag{7}$$

represented the best fit for the data on standard metabolism in micro-liter per hour at 20°C for organisms of wet weight W mg. He provides conversion factors for other temperatures but stresses that such relationships are designed to evaluate experimental results rather than replace them. *Blaxter* (1966), recalculating the data of *Holliday et al.* (1964) for herring, concluded that in larvae of mean wet weight 1.53 mg respiration amounted to 17 µl O_2/day at 8°C. Using Winberg's relationship a value of 23 µl O_2/day is arrived at, fair agreement considering his relationship is operating at its limit.

Deductions can be made about the type of metabolism involved in respiration. The consumption of 1000 l O_2, corresponds to the utilization of different materials as follows :

$$1000 \text{ µl } O_2 = \begin{array}{l} 1.05 \text{ mg protein} \\ \text{µ}1.23 \text{ mg carbohydrate} \\ 0.50 \text{ mg fat} \end{array}$$

1 mg of protein yields 4.25 cal, 1 mg of carbohydrate yields 4.15 cal, and 1 mg of fat yields 9.45 cal;

therefore,

$$1000\ \mathrm{l\ O2} = \quad\begin{array}{l}4.5\ \text{cal from protein}\\ 5.1\ \text{cal from carbohydrate}\\ \underline{4.7\ \text{cal from fat}}\\ 4.77\end{array}$$

The value 4.77 cal/1000 μl O$_2$ is sometimes called the oxycaloric coefficient. This makes it possible to calculate the requirements for growth and metabolism from measurements of growth rate and respiration. On the basis of 1000 μl O$_2$ being equivalent to 4.77 cal, and 1 mg wet weight of growth corresponding to 1 cal, *Ivlev* (1960b) concluded that young Baltic herring of wet weight 66.5 mg required 16.2 cal/day for growth and maintenance. These and other data are given in Table 4.6.

TABLE 4.7

Calorific Intake in Some Species

Species	Wet weight (mg)	Daily intake (cal)	Temp. (°C)	Author
Engraulis japonicus (anchovy)	1500-6300	78-426	14-20	Takahashi and Hatanaka (1960)
Alburnus alburnus (bleak)	250	38.3	19-22	Ivlev (1960b)
Clupea harengus (herring)	66.5	16.2	(?)15-17	Ivlev (1960b)
C. harengus (herring)	1.53	0.18	8	Blaxter (1966)

Smith (1957, 1958) reviews the work on growth, heat production, and respiration. High respiration and high rates of growth appear to be correlated, at least at some phases of development. In particular, specific growth rate and heat produced show very similar trends at various stages of development of the rainbow trout, *Salmo gairdneri (irideus)*. Up to hatching, plots of heat production on specific growth lie on a straight line with positive slope passing through the origin; after hatching the line passes through an intercept on the ordinate. There are two components in heat production — one resulting from growth and the other from maintenance. Temperature may affect the relative requirements of growth and maintenance thus giving differences in efficiency of development with temperature optima.

Growth

Growth is influenced in the early stages by the ratio between embryo and yolk weight, especially by the problems of efficient yolk utilization and the need to mobilize the nutrient reserves. Higher maintenance requirements per unit weight in small organisms, surface-volume ratios, change of diet with age, feeding and searching potential are also important. Growth has often been shown in terms of length but there is a need for uniformity, i.e., whether the caudal fin or fin fold is included, and length gives no information on the condition (fatness) of the organism or of the changes in body proportions with age; this is especially true in the yolk sac stage and at metamorphosis. Similarly, wet weights introduce the error of varying water content; thus, dry weight is the most satisfactory measure.

Examples of the relative quantities of yolk and embryo in a large and a small egg at different stages of development have already been given in Figure 4.5. The increase of weight of the embryo and decrease of the yolk tend to be logarithmic in nature. It is clear that in the earlier stages relatively less yolk is being used for maintenance than for growth, giving a higher efficiency of utilization. Without external feeding the larval weight begins to fall before all the yolk is resorbed. *Farris* (1959) divided the growth in length of four species of pelagic larvae after hatching into three phases: an early rapid phase after hatching, a slow phase near the completion of resorption, and a subsequent negative phase if no food was available. This shrinkage in terms of length was also described by *Lasker* (1964) in the larvae of *Sardinops caerulea*.

Growth of salmonid and other fresh-water fry in hatcheries has received considerable attention from an economic point of view, to provide maximum growth at minimum cost. Growth in marine conditions, where sampling of the same population is difficult owing to mortality, migration, or changes in net avoidance with age, is less adequately known, although examples are given in Figure 4.8.

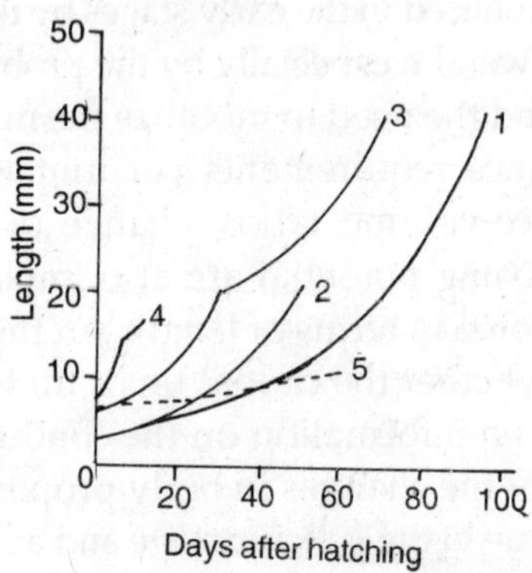

Fig. 4.8.   Length of different species related to age under natural conditions. 1. *Melanogrammus aeglefinus;* 2. *Scomber scombrus;* 3. *Clupea harengus;* 4. *Esox lucius* ; and 5. *Pleuronectes platessa.*

Some estimates of survival of fish larvae depend on a knowledge of growth rate and this is often inadequate or based on flimsy experimental data. Comparisons from tank experiments are difficult owing to mortalities, size-hierarchy effects, or other possible aquarium artifacts. Examples of size hierarchies in rearing experiments are given in Figure 4.9, showing the enormous range of size which may be found from an initially fairly uniform stock. Thus *Shelbourne* (1964), after a series of rearing experiments on plaice, found the length ranged from 7.5 and 37.5 mm. In dense populations there were also relatively larger numbers of small larvae. *Magnuson* (1962) tested the effect of food supply on *Oryzias* kept at high densities. As long as food was adequate, growth was not retarded, social hierarchies did not develop and aggressive behaviour was not manifested. Territorial behaviour only occurred when food was limited spatially. Whether size hierarchies occur in natural conditions is not known. It seems they must stem from inherently different growth rates or from some type of dominance hierarchy — the first could occur in the wild, but the second is more likely to occur in crowded tank conditions.

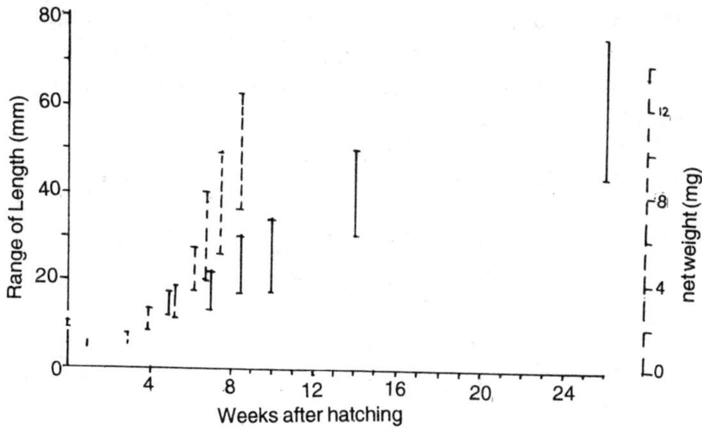

Fig.4.9. Size hierarchies in rearing experiments: (—) *Clupea harengus* length (left-hand ordinate), and (- - -) *Pleuronectes platessa* wet weight (right-hand ordinate).

Food supply and temperature are the most potent environmental factors in controlling growth in the later stages and there are few data available on this except for salmonids. In the egg stage low temperatures often produce longer larvae at hatching although *Lasker* (1964) found the maximum length of *Sardinops caerulea* larvae larvae at intermediate temperatures. There are then probably temperature optima for yolk utilization which partially give rise to these effects, but often the range of temperature used may be insufficient to show them. On the assumption that high temperatures are nonoptimal other conditions producing smaller larvae, such as low oxygen tensions and current velocities for salmonids as well as high salinities in the desert minnow, *Cyprinodon macularius,* shown by *Sweet* and *Kinne* (1964) and in cod, *Gadus macrocephalus,* by *Forrester* and *Alderdice* (1966), may also be considered nonoptimal. Combinations of temperature and salinity suggest that the optimum conditions for growth in length over the ranges used were 26°C for 70% salinity, 28°C for 35% and 33°C for freshwater in *Cyprinodon,* and 6° C and 19% for *Gadus macrocephalus.* The mortality, length, and maximum weight of salmon alevins (on the yolk supply) was greatest when they were reared in the dark, on a grooved surface, where presumably activity was minimal.

TABLE 4.8

Effect of Temperature on Size, Using Yolk Reserves Only

Species	*Method*	*Temp. range (°C)*	*Corresponding size range (mm)*
Salmo trutta (brown trout)	Maximum wet weight	5-16	135-95 (mg)
Coregonus clupeaformis (whitefish)	Hatching length	0.5-10	13-8.8
Osmerus eperlanus (smelt)	Hatching length	12-18	5.2-4.6
Cyprinodon macularius (desert minnow)	Hatching length	28-35	5.3-3.7
			4.2-3.6
Coregonus wartmanni (whitefish)	Hatching length	1-7	11-9
Gadus macrocephalus (Pacific cod)	Hatching length	2-10	4.1-3.5

Endocrines, Growth and Metamorphosis

Russian, Canadian, and other work on the role of the endocrine system in fish larvae is fully reviewed by *Pickford* and *Atz* (1957) and the role of the thyroid in metamorphosis by *Barrington* (1961). The main techniques used were histological, histochemical, or immersion in dilute solutions of thyroid hormone or antithyroid drugs. Evidence from a number of sources suggests that the pituitary is inactive until metamorphosis, for example, in herring *Clupea harengus*, eels *Anguilla anguilla*, bream *Abramis brama*, milkfish *Chanos chanos*, and sturgeon *Acipenser stellatus*.

The thyroid varies in its histological "activity" in the early larval stages but at metamorphosis in sturgeon, flatfish, eels, herring, pilchard *Sardina pilchardus*, and bone fish *Albula vulpes*, there are signs of secretion of colloid into the lumen of the thyroid follicles. Whether this is storage, or release into the circulation is taking place, can be satisfactorily tested only by measuring hormone levels in the blood. Earlier, thyrozine may retard growth and accelerate morphogenesis (e.g., in salmon, sturgeon, *Misgurnus fossilis* and *Lebistes*) at a concentration of 1 ppm or less. Rate of oxygen consumption may also be increased. In the developing eggs of *Scyliorhinus canicula* (see *Dimond*, 1963) the functioning of the thyroid seems highly dependent on temperature. At 8°C use of ^{131}I showed the thyroid concentrated and bound iodine, but there was no evidence of thyroid hormone formation. At room temperature thyroid hormone reached a measurable level. Thiourea at concentrations of about 100 ppm may delay hatching, retard yolk resorption, and inhibit growth in salmonids, inhibit morphogenesis in sturgeon larvae, and decrease oxygen consumption is salmon and sturgeon larvae by as much as 15%. Thiourea seems to improve the ability to utilize oxygen at low tensions, although it is not clear whether this is because of its antithyroid action. Thiourea might well improve survival in suboptimal conditions of oxygen concentration that might occur, for example, during transportation. The effect of thyroxine on promoting metamorphosis was tested on the eel (*Vilter*, 1946). There was an accelerating, but not a sudden, effect with an indication that an increasing threshold controls the sequence of events in metamorphosis. In other words, the later stages of metamorphosis require higher-concentrations of thyroxine. Other experiments on lampreys showed no effect of thyroxine, thyroid extract, iodide, or iodine on metamorphosis (see *Barrington*, 1961).

It may be concluded that there is a suggestion of the thyroid playing a role in growth and differentiation in the larval stages but further work is required, especially at the important metamorphosis stage.   If the pituitary is not functioning until metamorphosis the thyroid must be relatively independent of pituitary control in the early stages of life.

FEEDING, DIGESTION AND STARVATION

Often the mouth is not completely formed at hatching, but rapid development in many marine fish larvae leads to the possibility of taking external food before the yolk is finally resorbed. Without success in feeding there is eventually self-metabolism and loss of weight. Yolk may be transported and stored within the body, especially in the base of the primordial fin and other subdermal spaces, and fat may be stored in the mesenteries of the larval gut. Unlike demersal fresh-water species there is no vitelline circulation in most pelagic larvae although *Scomberesox*, *Trachypterus*, and a few other species are known to be exceptions. Usually there is a yolk sac sinus in connection with the heart and with lateral branches to the subdermal spaces; in the species with a vascularized yolk sac these spaces are much less inflated.

After final resorption of the yolk the larvae retain their potential to feed for some days depending on species, egg size and temperature. The concept of a "point of no return" was introduced by *Blaxter* and *Hempel* (1963) for herring larvae. This point, after which the larvae are still living but too weak to feed, may vary from 15 days after fertilization for summer spawners in warmer water to 45 days for winter spawners. In the larval cisco, *Leuchichthys artedi*, it was 27-36 days after hatching at 3°-4°C. The activity of larval *Solea solea* drops from about 70% of the time active at first feeding to 20% some 5-6 days later when, if feeding is unsuccessful, inanition occurs. Information on the time to point of no return in relation to spawning and food supply may help to predict the probability of survival in different broods.

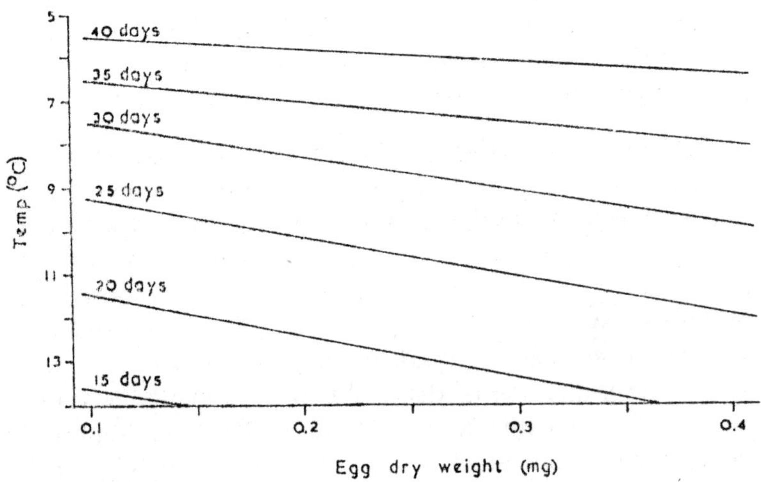

Fig. 4.10.   The concept of "point of no return." The time from fertilization to the point where the larvae of *Clupea harengus* are too weak to feed is given in the form of a nomogram based on original egg dry weight of different races and the temperature.

Most fish larvae are predatory with a large mouth and well-developed eyes. In herring the gape of the jaw increases by 50% during the yolk sac stage: the gape of 0.3-0.4 mm at first feeding depends on the length of the larvae and, therefore to some extent, on the original egg size. An elastic ligament at the articulation makes it possible for even larger organisms to be taken. Some species have very small gapes which is partly a function of their small size (e.g., lemon sole *Microstomus kitt*, sprat *Sprattus sprattus*, pilchard *Sardina pilchardus*, sardine *Sardinops caerulea*, and many gadoids). These must require very small food organisms for their early survival, a factor which is of considerably less importance for many fresh-water fish.

The forward darting movements of fish larvae in order to engulf the prey are described in a number of species, for instance, in *Coregonus* and *Esox* by *Braum* (1964) and in sole,

Solea solea, by *Rosenthal* (1966). *Iwai* (1964) found three types of feeding in the larvae of *Plecoglossus altivelis:* a predatory snapping, a respiratory current with a sieving action by the gill rakers, and a ciliary current associated with the olfactory pits.

The extent of which fish larvae are herbivorous and might make use of such currents is not clear. Green food in the gut may be incidental to the swallowing of water or due to the fecal material from prey species, and its presence is no evidence of utilization. In the anchovy, *Engraulis anchoita*, however, phytoplankton in the guts of young stages can be correlated with the formation of gill rakers which presumably act as a sieve. The utilization of dissolved organic matter has also never been adequately tested. *Morris* (1955), in a reappraisal of the theory, concluded from circumstantial evidence of a low incidence of feeding in larvae collected from nature, that it was likely they were utilizing at least fine particulate matter, which could be collected by mucous cells within the buccal cavity or during swallowing of seawater for osmoregulatory purposes. Morris estimated that the larvae of the night smelt, *Spirinchus starksi*, might pass about 1.15 ml of water per day across these mucous surfaces, possibly containing 0.011 mg of organic matter, a very small quantity even for a larva of 0.9 mg wet weight. The recent work of *Terner* (1968), mentioned in Section IV, D, showing that the ova of *Salmo gairdneri* can probably incorporate exogenous substrates into their endogenous reserves, may also be significant.

The guts of many pelagic marine larvae are often empty, a puzzling feature when one considers the fairly high food requirements in the early stages. Apart from a lack of suitable food this may result from the capture by net of the weaker feeders only, rapid digestion, defecation during capture, or diurnal feeding rhythms. Examples of such rhythms, which are given in Figure 4.11, do indicate a decrease in feeding at night and an increase at dawn.

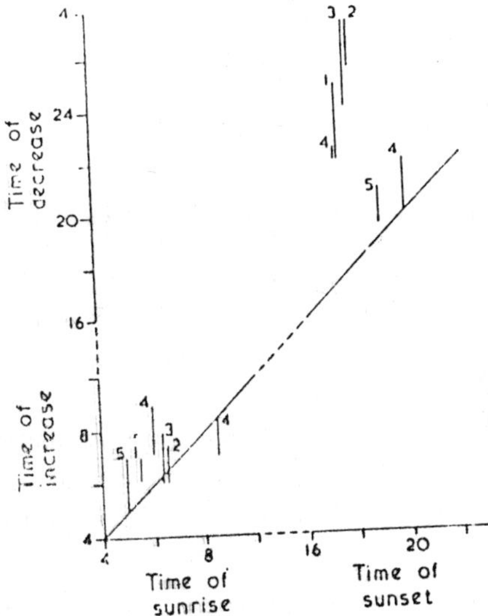

Fig. 4.11. The time period over which the gut contents of various species of larvae increase and decrease related to the time of sunrise and sunset. Note the lag at sunset caused, presumably, by the time taken to digest food taken earlier. 1. *Pleuronectes platessa*; 2. *Pleuronectes platessa*; 3. *Ammodytes* ; 4. *Clupea harengus*; and 5. *Salmo salar parr*.

The role of vision in feeding has been tested experimentally and thresholds of light intensity measured. The fall off in rate of feeding corresponds with the dusk and dawn periods. Some species can take food in the dark, e.g., cisco, especially when it is present in high concentrations. Some, like sole, take food in the dark for most of the larval life and others, like plaice, only at later stages around metamorphosis. Where vision is important the daily feeding period must vary considerably with season, latitude, clarity of water, and even average cloud cover. *Ivlev* (1960b) calculated, from the food requirements of young Baltic herring, their rate of feeding and the density of food, that they needed to feed 15 hr/day in August, in other words nearly all the hours of daylight. *Blaxter* (1966) estimated that much younger stages of herring had 10 hr/day to feed in the southern winter spawning groups and up to 24 hr/day in the more northern summer spawners. Variations in feeding time affect searching power, and therefore survival and growth, and these may be further influenced by the temperatures prevailing in different seasons.

TABLE 4.9

Main Range of Light Intensity over which Feeding Becomes Reduced

Species	Range of light intensity (me)	Author
Oncorhynchus keta (chum salmon)	10^1-100^-	Ali (1959)
O. gorbuscha (pink salmon)	10^1-10^{-3}	Ali (1959)
O. nerka (sockeye salmon)	10^1-10^{-3}	Ali (1959)
O. kisutch (coho salmon)	10^0-10^{-4}	Ali (1959), Brett and Groot (1963)
Coregonus warimanni (whitefish)	?-10^0	Braum (1964)
Esox lucius (pike)	?-10^{-1}	Braum (1964)
Clupea harengus (herring)	10^2-10^{-1}	Blaxter (1966)
Pleuronectes platessa	10^2-10^{-2a}	Blaxter (1968b)

[a]Older stages feed in the dark.

Success in early feeding, feeding drives, and learning factors have been studied. At very early stages *Braum* (1964) found only 3-8% of feeding movements in *coregonus* were successful, but 30% in *Esox*. In the yolk sac stage, herring larvae take food successfully in 3-10% of their feeding movements but later in 80-90%. In plaice, feeding is 60-80% successful throughout larval development. In herring larvae from 20 to 40 mm in length the feeding drive depends on factors such as satiation and especially on the activity of the prey. Live organisms, although more difficult to catch, seem to increase the feeding drive and are probably very important in the young stages, where the main visual process may centre round movement perception. Larval cisco, however, showed a much higher rate of feeding on dead than live *Cyclops*.

The gut is usually a straight or simple tube at first feeding and the food is often digested near the anus. This is an interesting reverse of the normal situation in adult vertebrates. It seems likely that the gut is unspecialized at this stage and digestive enzymes are secreted along its length, although no critical work has been done on this. *Harder* (1960) gives a detailed account of the changes in the gut during the early growth of clupeids and engraulids; *Ryland* (1966) gives a rather simpler account for the plaice. Movement along the gut is mainly by peristalsis, although *Iwai* (1964) found cilia in the gut of ayu, *Plecoglossus altivelis*, especially in positions posterior to the liver. Backwardly directed ciliary currents and forward peristalsis seemed to cause a circulation of the gut contents, perhaps to aid digestion. In further reports use of the electron microscope showed intermingling of ciliated cells and columnar cells with microvilli in both *Plecoglossus* and *Hypomesus olidus*. Reduction of cilia in later stages suggested a transition from cilia to microvilli. Whether cilia are to be found in many species is still open to investigation.

Rates of digestion have been measured by rate of disappearance of the gut contents (the gut wall and body being transparent) or by interposing differently coloured food in the normal diet and observing the first signs of coloured feces. *Kurata* (1959), using herring 12 days old, found that the gut clearance rate ranged from 12 to 19 hrs or more at 9°C depending on the extent of food intake. The rate of digestion varies from 9 hrs at 7°C to 4 hrs at 15°C judged by transparency of the gut contents in this species.

Reports of larvae in poor conditions or dead in net hauls may be due to selective capture or poor washing and are not certain evidence for mortality. *Shelbourne* (1957) estimated the condition of plaice larvae in the North Sea and found more signs of inanition early in the year. Measurements of condition

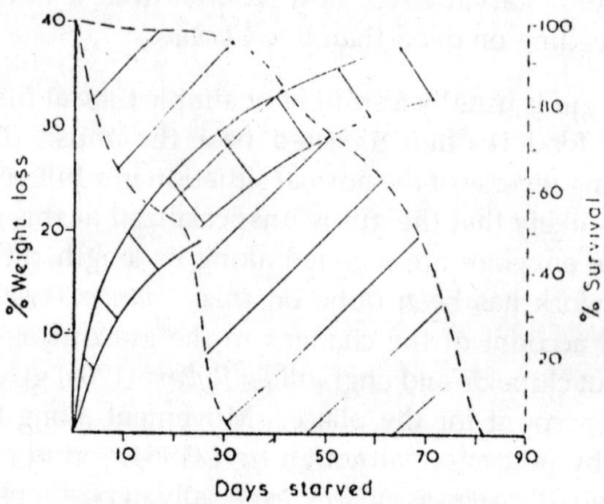

Fig. 4.12.   Weight loss (—and left-hand ordinate) and per cent survival (- - - and right-hand ordinate) as envelopes" for 12 species of fish larvae during starvation.

factor [weight/length^3 x 1000] in herring larvae compared with starving larvae in tanks suggested that larvae in the sea were often very near the point of starvation although, over a number of years, it was not possible to correlate high condition factors with a good food supply. *Ivlev* (1961) looked at various aspects of starvation in young catfish, *Silurus glanis* and bream, *Abramis brama*.   With complete starvation the survival time was 34-46 days corresponding to a weight loss of 33-39%. With rations below the maintenance requirement survival was prolonged, but even on a full maintenance diet, the weight remaining constant, the fry died after 126-151 days, indicating that a "need" for growth may be characteristic of these young stages and that merely to maintain uniform weight is physiologically or developmentally inadequate. The current velocity stemmed by young carp *Cyprinus carpio* and roach *Rutilus rutilus* for 5 min was reduced by 10 times, from about 100 to 10 cm/sec, over 52 days of starvation.   Age had a considerable influence on survival time without food.   In catfish, roach, and bleak *Alburnus alburnus*, it ranged from 3 to

6 days at yolk resorption and from 109 to more than 180 days
when 100 days old. The loss in weight and percentage survival
of 12 species of larvae 25-30 days old during stravation is
shown in Figure 12. Losses of weight of 30-35% were possible
before death, with survival times of 30-80 days. Starving fish
were much more susceptible to unfavorable conditions and
also to infection and predation.

TABLE 4.10

**Lethal Limits of Certain Environmental Factors on Fry Before and After
Starvation**

Species •	Normal limit(s)	Starvation limit(s)	Days starved
	Salinity %		
Alburnus alburnus (bleak)	4.5	2.5	20
Caspialosa volgensis (Volga herring)	16.5	14.5	20
	pH		
Tinca tinca (tench)	4.0-1.8	5.7-8.3	50
Perca fluviatilis (perch)	5.3-8.2	6.3-7.8	30
	O_2 ppm		
Cyprinus carpio (carp)	0.7	2.1	50
Alburnus alburnus (bleak)	2.8	3.1	20

SENSE ORGANS

A. Vision

The newly hatched larvae of many species have
unpigmented, and presumably nonfunctioning, eyes (e.g., sole
Solea solea, mackerel *Scomber scombrus*, whiting *Merlangius
merlangus*, pilchard *Sardina pilchardus*, and sardine *Sardinops
caerulea*); others have pigmented eyes (e.g., plaice *Pleuronectes
platessa*, cod *Gadus morhua*, herring *Clupea harengus*, and

salmonids). *Schwassmann* (1965) examined sections of the eye and brain of larval *Sardinops caerulea* after hatching. On the first day there was no pigment and little differentiation, but by the second day pigment with a stratified retina and visual cells had developed. At 5 days the optic nerve was myelinated and the optic chiasma showed some interdigitation. Considering there are only about 40 individual fibers in the optic nerve when feeding commences at 3 days, it seems probable that the eye at first feeding is only capable of coarse movement perception. The larval eye is also limited in other respects. In newly hatched and *Oncorhynchus* and in *Clupea harengus* and *Pleuronectes platessa* and in a number of other species up to metamorphosis, there is a pure-cone retina, although the adult has both cones and rods. *Lyall* (1957a, b) examined the developing eye of the trout, *Salmo trutta,* and concluded that the rods might develop from single cones or by an outward migration of bipolar cells, the latter view being supported by *Blaxter* and *Jones* (1967).

Retinal pigment migration and other retinomotor responses associated with dark and light adaptation only develop when the rods appear and are therefore absent in herring and plaice until the larvae metamorphose. This, may be true of many other marine species. Thus, the larvae are, visually, poorly equipped, with a single type of visual cell, no ability to dark or light adapt, and with a coarse retinal mosaic. *Oncorhynchus* species acquire rods and retinomotor responses very rapidly after hatching compared with the marine fish which have been examined. The range of light intensity over which light and dark adaptation takes place is then 10^1-10^{-1} meter-candles (mc), which corresponds fairly well with the thresholds for feeding.

The acuity of the larvae is poor compared with the adult. Although the retinal cells are small and fairly closely packed, the eye is also small, as is the focal length of the lens. *Baerends et al.* (1960) found that young *cichlids, Aequidens portalegrensis,* could be trained to distinguish stripes 1.5 mm apart at 3 cm body length; this improved to 0.3 mm at 11 cm body length. *Blaxter* and *Jones* (1967) calculated the acuity of the eyes of larval herring (minimum separable angle) to be 200 minutes at 1 cm body length improving to 50 minutes at metamorphosis.

Thresholds and spectral sensitivity were measured by behaviour techniques in herring, plaice, and sole larvae. Using a negative phototaxis as criterion the visual thresholds were about 10^{-5}-10^{-6} mc, the spectral sensitivity curves being plateaulike with peaks which might correspond to different cone populations. Distances of perception and visual fields have been measured by a number of workers. These are given in Table 4.10.

B. Neuromast Organs

It has probably not been fully appreciated that many species of fish larvae have free neuromast organs. Iwai reviews earlier reports and his own work on *Tribolodon hakonensis*, *Tridentiger trigonocephalus*, and *Oryzias latipes* (1967a). These organs are usually situated in lateral rows along the body with the small hump of sensory cells being the main element visible especially in sea-caught or fixed specimens. Rearing, with the use of phase contrast microscopy, makes it possible to retain and see the cupulae which are relatively enormous — 0.05 mm long in *Oryzias* but 0.1-0.4 mm long in some other species. In the adult it is likely that they are more often sunk into pits. The ability of larvae to avoid capture before the eyes fully develop is almost certainly a result of these neuromast organs.

ACTIVITY AND DISTRIBUTION

A. Phototaixs and Activity

Much of the published work on phototaxis in fish larvae is aneodotal. Generalizations that marine fish larvae are photopositive and demersal fresh-water larvae photonegative do not stand up to a detailed investigation using different temperatures, light intensities, species, or ages. Changes of the sign of the phototaxis may be related to temperature, especially where a sudden change is applied, e.g., in Coregonus. Herring larvae show a positive phototaxis at high light intensities which becomes negative below a certain threshold.

TABLE 4.11

Visual Fields and Perception Distances

Species	Size	Distance for locating (mm)	Feeding distance (mm)	Visual field Single eye	Binocular
Coregonus (whitefish)	Early larvae	10	0.5-3	145°	45°
Esox lucius (pike)	Early larvae	(?)10	3-4	150°	
Clupea harengus (herring)	12-14 mm	5	3		
Clupea harengus (herring)	10-19 mm	10-26			
Pleuronectes platessa (plaice)	6-8 mm	5	<5		

Activity levels are frequently associated with diurnal light changes and phototaxis. *Hoar* (1956, 1958). *Ali* (1959), and *Heard* (1964) review much work on activity, rheotaxis, schooling and migration of young salmonids. Emergence of the alevins of sockeye salmon. *Oncorhynchus nerka*, seems to be controlled by incident light on the gravel and can be delayed by artificial light at night. The younger fry of pink salmon, *Oncorhynchus gorbuscha*, for example, are negatively phototactic and hide among the stones of their home stream by day. They become active at night and may rise to the surface and swim downstream or get displaced by the current. After schooling both pink and chum salmon *O. keta*, prefer lighted conditions and presumably substitute schooling for hiding as their protection against predators. They show then a positive phototaxis, stem the current by day displaying a cover reaction only with abrupt changes of light, and again become displaced downstream at night when the visually controlled rheotaxis phases out. Coho salmon fry, *O. kisutch*, are rather different in behaviour; they are strongly territorial and generally more active, but their responses to light and their diurnal changes in activity are much less marked. The behaviour of the fry of the different species of *Oncorhynchus* can, in fact, be related rather generally to their migratory habits. Pink and chum fry migrate to the sea quite soon after hatching, whereas sockeye fry tend to move into lakes, and coho remain in the home stream for a year or so. Loch trout, *Salmo trutta*, also show incipient territorial behavior at an early age. *Woodhead* (1957), in a laboratory study of the larvae of *Salmo* species, related changes in photokinetic activity levels to ambient light intensity and age. Between 0.005 and 100 mc the photokinetic activity of *S. trutta* gradually increased above a basal dark level, and this species and *S. gairdneri (irideus) and S. salar* all showed increased activity with age. A positive photokinesis, together with a negative phototaxis, would work together to keep the larvae within the stones of their home stream. *Woodhead* and *Woodhead* (1955) found a similar increase in the activity of herring larvae with light intensity above 0.3 mc, and there was a basal level of activity in very low illumination as there was in the salmonids.

B. Vertical Distribution

Diurnal variations in distribution are linked to activity, phototaxis, and brightness discrimination. Some earlier work on changes of vertical distribution did not take into account the need for opening and closing the plankton nets at defined depths and for preventing differential net avoidance by day and night. A general lack of larvae by day, especially larger specimens, partly results from this latter factor, as emphasized by *Bridger* (1956). With slow two nets it can be difficult to estimate the relative effects of net avoidance and vertical distribution when comparing day and night catches. *Strasburg* (1960) used slow tow nets to catch tuna larvae. Some species such as skipjack, *Katsuwonus pelamis*, were rarely caught by day at all while catches of all species were always much higher by night. He concluded that there was some net avoidance, but the main diurnal difference was owing to vertical migration. Stevenson (1962) used high speed nets for larvae of Pacific herring, *Clupea pallasii*. The catches at night were far higher and the average size of the larvae was greater, from which he concluded that the larvae were concentrated at the surface by night and spread over a fairly wide stratum by day. There seemed to be net avoidance, even of high speed nets, by the larger larvae in the daytime. *Ryland* (1964) found a reverse distribution of larval plaice and sand eels *Ammodytes*, which were spread over a wide depth range from the surface to about 35 m by night and much more concentrated between 5 and 10 m during the day. The use of high speed nets in this study and in that of *Colton* (1965) on haddock larvae, *Melanogrammus aeglefinus*, prevented diurnal variation in size and presumably overcame the net avoidance factor.

C. Buoyancy and Pressure

There are a number of devices which provide buoyancy in pelagic fish eggs and larvae. In marine species, found in high salinities, maintenance of the body fluid concentration below that of the environment acts as a buoyancy mechanism. In the early stages lack of a skeleton also keeps the specific gravity fairly low. In fresh-water the eggs are often demersal so that the much greater problem of maintaining buoyancy does not exist. In fact, in salmonid eggs the specific gravity may be very high in the early stages because of low water content of the yolk. It is sometimes said that a large perivitelline space is a mechanism for buoyancy. However, this normally contains water of the same concentration as the environment and is thus much more likely to act as a shock absorber, particularly in pelagic fish eggs subject to wave action. The embryo and yolk may regulate the body fluids from fertilization (e.g., in pelagic plaice eggs: by means of the vitelline membrane. In the demersal herring egg there is no regulation until the yolk is covered by cellular tissue at gastrulation. Many species such as pilchard *Sardina pilchardus*, mackerel *Scomber scombrus*, hake *Merluccius merluccius* and sole *Solea solea* have oil droplets, but these are not typical of pelagic eggs for they are absent in the sprat *Sprattus sprattus* and in *Gadus* and *Pleuronectes* species. Tendril-like outgrowths found, for example, in *Exocoetus* and Scomberesox eggs are as likely for attachment as for buoyancy.

There may be a mechanism to adjust specific gravity to the spawning medium. *Solemdal* (1967) found that variations in the osmotic pressure of the blood of female flounders. *Pleuronectes flesus*, at the spawning season depended on the salinity. This, in turn, caused changes in the specific gravity of the eggs when spawned. In water of low salinity the eggs were larger and their osmotic pressure lower. In the low saline water of the Baltic (6.5%), fertilization and development were possible but the eggs rested on the bottom. In the sprat there is a correlation between low salinity and larger eggs and also in herring and plaice.

In the larval stages large subdermal spaces as in the plaice and cod will reduce specific gravity as long as the body fluids are maintained at an osmotic pressure below that of the environment. Other species have long processes to the dorsal or lateral fins, e.g., the angler fish *Lophius* and deal fish *Trachypterus*. It may be significant that *Orton* (1957) quotes the latter species as having only moderate fin folds and subdermal spaces, but whether the increased surface area assists buoyancy is open to question; these may be organs to reduce predation. In fresh-water, the larvae of salmonids can afford to retain dense yolk reserves as they rest on the bottom. Others like carp and bream have cement organs for attaching themselves to weed.

The first filling of the swim bladder is well known in fresh-water fish. In the physostomatous salmonids like *S. trutta, S. gairdneri, Cristivomer namaycush,* and *Coregonus clupeaformis* the swim bladder is filled 1-3 months after hatching, depending on temperature or feeding. Air is taken in from the surface via the pneumatic duct and without access to the surface the swim bladder remains empty. The pneumatic duct remains open under such conditions and functions later. Lake trout, *Cristivomer namaycush*, were able to swim considerable vertical distances (280 m) with unfilled bladders suggesting that development in deep water presented no problems for filling the swim bladder. Physoclists like *Gasterosteus*, *Lebistes* and *Hippocampus* also require access to air to fill the swim bladder but if deprived of this at the critical time the pneumatic duct closes and the swim bladder cannot be filled later.

Pressure effects on fish larvae were tested briefly by *Bishai* (1961). Newly hatched herring survived compression to and decompression from 4 atm and young plaice 37-50 mm long to and from 2 atm. Young salmonids could live at 5 atm to the end of the yolk sac stage, sometimes showing increased activity during compression or decompression. Older fry, about 20 days after yolk resorption, could withstand compression to 2 atm but had difficulty in dealing with air bubbles inside the body and in the swim bladder during decompression. Older fry still, about 60-140 days after yolk resorption, could not even withstand compression to 1.2 atm. At later stages, the ability to withstand compression increased but decompression was lethal. *Qasim et al.*, (1963) observed changes in the vertical distribution of larval teleosts subjected to pressure changes. Young plaice larvae tended to swim upwards when the pressure was increased from 1 atm to 2 atm and sank when the pressure was returned to normal. Metamorphosing plaice did not show this response. Larvae of the blenny *Centronotus gunnellus* responded in a similar way to pressure changes equivalent to 25 cm of seawater and larvae of another blenny *Blennius pholis* to only 5 cm of seawater. Only the last of these species has a swim bladder. It is possible in such experiments that the larvae could have been responding to water currents produced by the pressure changes. If this is not so, one may ask how larvae without swim bladders can respond to pressure.

D. Locomotion and Schooling

Most species show signs of activity within the egg, and movement is often an important part of the hatching process.

The relatively large yolk sac at hatching must be a hydrodynamic embarrassment to the larvae, but in salmonids, where it is especially large, there is little movement as the larvae grow in a rather passive way within the stream bed. The buoyant oil globule found in many marine pelagic larvae may also make equilibrium difficult. Two sorts of movement are found at yolk resorption — a serpentine eel-like mode of swimming in the long thin-bodied clupeoid type and a more maneuverable movement allowing "backing-up" with the use of the pectoral fin and marginal fin folds in the shorter flatfish type of larvae.

Swimming performance is shown in Figure 4.13 in terms of burst speeds maintained for a few seconds, where the theoretical 10 body lengths/sec (shown as dashed line) seems to fit quite well. Cruising speeds, of the order of 2-3 body lengths/sec (ideally the maximum sustainable speed without an oxygen debt accumulating), are shown in Table 4.11. Improvements in performance can be correlated with development and upturning of the caudal fin.

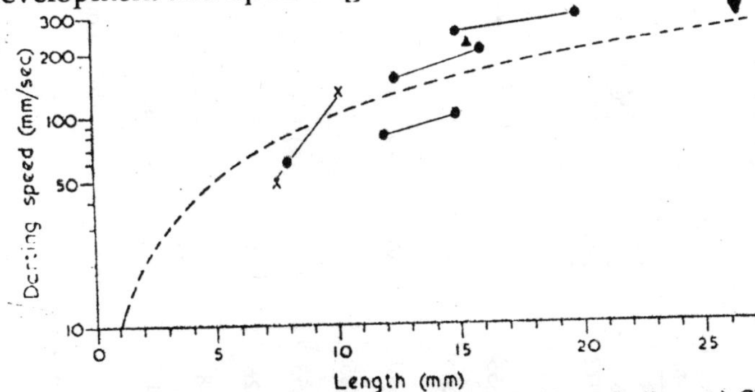

Fig. 4.13. Darting speeds of larvae of different length: (●——●) *Clupea harengus* 10°-15°C (X——X) *Pleuronectes platessa* (6.5o-7.5° (▲) *Carassius auratus*, no temperature and (▼) *Perca fluviatilis*, no temperature (Radakov, 1964).

If schooling occurs it usually starts at metamorphosis. Presumably the larvae, being transparent, do not provide very good mutual stimuli for keeping schools together visually. *Shaw* (1960, 1961) analyzed the development of schooling in

TABLE 4.12
Cruising Speeds of Some Young Fish

Species	Length (mm)	Speed (mm/sec)	Temp. (oC)	Time maintained	Distance swim (m/hr)	Author
Pleuronccles platessa (plaice)	7.5	12	6.5-7.5	A few sec		Ryland (1963)
	9.6	27	6.5-7.5	A few sec		
Coregonus teartmanni (whitefish)	10-12	16	4	?	58	Braum (1964)
		29	16	?	103	
Rutilus rutilus (roach)	35-45	138	?	1 min		Aslanova quoted by Radakov (1964)
Abranis brama (bream)	45-55	126	?	1 min		
Cyprinus carpio (carp)	50-60	129	?	1 min		
Teachurus trachurus (horse mackered)	30-40	136	?	1 min		
Mullun barbutus (mullet)	35-45	128	?	1 min		
Mugil sp. (mullet)	45-55	156	?	1 min		
Blicea bjorkna (white bream)	18-26	330	?	?		Radakov (1964)
Clupea harengus (herring)	6.5-8	5.8	(?14)	45 min		Bishai (1960d)
Clupea harengus (herring)	12-14	10	9-10	60 min	36	Blaxter (1966)
Solea solea (sole)	4	6-9	15	Long periods intermittent swimming	17-19	Rosenthal (1966)
Alburuns alburnus (bleak)	0.26g (wet wt)	29.8	20		107	Ivel (1960b)
Micropterus dolomieui (smallmouth bass)	22	146-312	20-35			Larimore and Duever (1968)

the silverside *Menidia* and found responses initiated between two fry at a body length of 8-9 mm with increasing participation as they grew. By 11-12 mm as many as 10 could be seen in parallel orientation. Fish reared in isolation in paraffin-coated bowls took a few hours to become integrated into existing school. Generally speaking there was a considerable latency in schooling when "isolates" and "quasi-isolates" (isolated 5-7 days after hatching) were brought together at a size where schooling was normally well-developed. Surprisingly, this initial latency was less among "isolates" than "quasi-isolates." In the newly emerged fry of *Oncorhynchus* species, schooling ceased in the dark and following subjecting to light they needed 15 min or more to complete the school again, although in all species except coho this time became reduced with age. In herring schooling started about 2 months after hatching at a body length of 25-30 mm. Starving larvae tended to school more positively, but the school rapidly broke up up when food was offered.

Rheotropic (optomotor) responses (orientation to currents or moving backgrounds) are often well-developed in the larval stage as shown in some experiments on swimming performance and from the many observations on salmonid larvae by day and night. This response seems to be maintained by either visual or tactile cues, and the sign may change under shock treatment. In young salmonids a temperature rise of 12°C changed the rheotaxis from positive to negative, presumably a defensive response.

E. Searching Ability

The ability to search for prey or, in the case of visual feeders, "optically filter" the environment, depends on cruising speed and the distance at which food is perceived. Perception distances depend on the movement of the prey, its orientation in relation to the eye, contrast with the background, and illumination. Activity in terms of cruising speed may well be reduced after a period of feeding and is certainly influenced by feeding drives in older larvae. Starvation may also cause a reduction in activity. The volume searched per hour has been estimated in whitefish and herring larvae. When multiplied up on a daily basis the volume searched becomes very dependent on day length, that is, on latitudinal and seasonal effects.

MORTALITY, TOLERANCE AND OPTIMA

From hatchery, rearing, and experimental studies it is well-known that fish eggs and larvae go through periods of mortality. Susceptibility to mechanical shock or nonoptimal temperatures and salinities may vary with age. Marine fish eggs seem especially delicate until the completion of gastrulation although this has not been systematically tested. Perhaps sensitive morphogenetic processes in the early stage or failure to osmoregulate are the cause. Similarly there are stages, such as the eyed stage in salmonids, where eggs may be particularly resistant to damage. After hatching, further stages of mortality may be experienced which can sometimes by connected with the further development of certain organs and with such physiological changes as the transition from cutaneous to gill respiration or the development of the swim bladder. Rearing studies on marine fish show a high mortality at the end of the yolk sac stage in species such as cod *G. morhua*, haddock *Melanogrammus aeglfinus*, pilchard *Sardina pilchardus*, herring *Clupea harengus*, and lemon sole *Microstomus kitt*. These "critical phases" are probably in part the result of unsuitable food being supplied, especially from the aspect of size. The question of whether critical phases at yolk resorption occur under natural conditions is a much vexed question, examined especially by fishery biologists looking for the factors controlling brood survival. *J.C. Marr* (1956), in reviewing this hypothesis, concluded in two species (mackerel *Scomber scombrus*, and Pacific sardine *Sardinops caerulea*) that mortality was more likely to be constant or steadily decreasing without the catastrophic periods advocated by earlier workers. Undoubtedly some of the five species mentioned have small mouths and the finding of suitable food in the sea at the end of the yolk sac stage before the point of no return must be a serious problem.

TABLE 4.13

Volume Searched during Feeding

Species	Size (mm)	Volume searched (liter/hr)	Author
Coregonus wartmanni (whitefish)	(?)10	14.6	Braum (1964)
Clupea harengus (herring)	8-16	0.3-2.0	Blaxter (1966), Blaxter and Staines (1960a)
Clupea harengus (herring)	10 13-14	1.5-2 6-8	Rosenthal and Hempel (1968)
Sardina pilchardus (pilchard)	5-7	0.1-0.2	Blaxter and Staines (1969a)
Pleuronectes platessa (plaice)	6-10	0.1-1.8	Blaxter and Staines (1969a)

The lethal levels of various environmental factors such as salinity, temperature, oxygen tension, pH, radiation and mechanical stimuli have been assessed for fish eggs and larvae. Salinity tolerance is surprisingly wide, ranging in marine species which have been examined from about 5%° up to 45-50%° with even a wider range over short periods of exposure. Predictably, the lethal temperature varies with level of adaptation. A wide variety of criteria have, unfortunately, been used for defining lethal temperature, ranging from 50% mortality after one day (or longer) to the number of days to 50% mortality. Examples of high and low lethal temperatures where the criteria could be standardized are given in Figure 4.14 related to temperature of acclimation. Temperature fluctuations are much damped down in the sea and may be negligible in deep water. In the tropics the seasonal fluctuations may also be slight, and this may lead to a narrow range of tolerance. In 10 species of fish larvae from the Indian Ocean, *Kuthalingam* (1959) found the range from lower to upper lethal

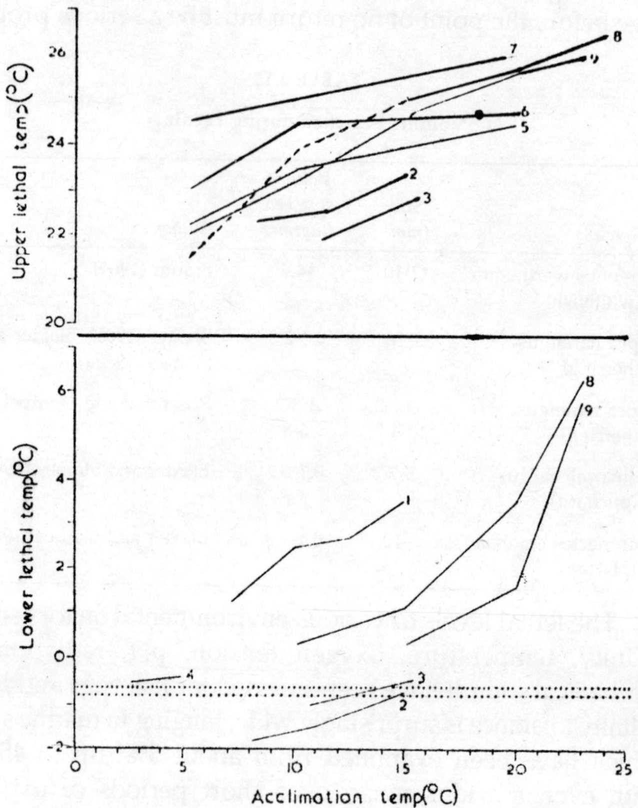

Fig. 4.14. Lower and upper lethal temperatures at different acclimation temperatures, criterion of lethal temperature being 50% survival after 24 hr; horizontal dashed lines give range of blood freezing point values. 1. *Brevoortia tyrannus* — 24% salinity, acclimated for 12 + hr. 2. *Clupea harengus* (spring spawned) and 3. *C. harengus* (autumn spawned)—34% salinity, acclimated for long period. 4. *Oncorhynchus keta*—in 28% salinity, acclimated for 21 days. 5. *Salmo salar*, 6. *S. trutta*, and 7. *S. trutta (fario)*— in fresh-water, acclimated for 5 days. 8. *Oncorhynchus tshawytscha*, and 9. *O. nerka*.

temperature was only 3°-4°C (about 28°-32°C). Temperature tolerance may also vary with age. *Combs* (1965) measured long-term temperature thresholds over the period from fertilization to hatching in *Oncorhynchus* species. In *O. nerka*, for example, the low temperature threshold was about 5°C and the high temperature threshold about 13.5°C. At the 128-cell stage and later both *O. nerka* and *Chinook, O. tshawytshca*, had a greater range of tolerances. In fresh-water or high latitude resistance to cold must be adequate or else adaptations are developed which prevent contact with extremely cold water. The flounder, *Pseudopleuronectes americanus*, is one of the few species of flatfish with a demersal egg and it spawns inshore in very cold water. Temperature optima may be judged by hatching success. For instance in *Coregonus clupeaformis* the highest percentage hatching was at 0.5°C with a very rapid fall off above 6°C. In herring temperature optima vary with the race. In medaka *Oryzias* the best survival was obtained when the rearing temperature was alternated between 22° and 30°C every 12 hrs . Generally, outside the optima not only does the percentage hatch fall but there are more abnormalities.

Oxygen is apparently only limiting at its low level. However, super-saturated solutions may be harmful if air bubbles are swallowed by the larvae. This "gas disease" can be fatal if the larvae cannot eliminate the bubbles from the gut as their buoyancy is disturbed. The eggs of the salmonids *S. salar*, rainbow trout *S. gairdneri*, Pacific salmon *O. tshawytscha* and *O. kisutch*, brook trout *Salvelinus fontinalis*, and lake trout *Salvelinus* namaycush, have been tested at low oxygen tensions. Most were able to resist levels of 2.4 ppm or even less (25% saturation) although development was retarded and the larvae smaller at hatching. Reductions in current flow had a similar effect. *Alderdice et al.* (1958), using chum salmon eggs, measured the retardation in hatching resulting from low dissolved oxygen levels at different developmental stages. The intermediate stages, about one-third of the way from fertilization to hatching, were most susceptible. *Hayes et al.* defined the limiting tension not from its lethal aspect, but the tension at which normal oxygen consumption started to drop. In *S. salar* eggs this was

3.0 ppm at 20 days, 7.5 ppm at 45 days, and 4.7 ppm at hatching.   Other salmonid data on limiting tensions are summarized by *Wicket* (1954). In the egg of chum salmon, *O. keta*, the calculated limiting tension varied from 0.72 to 3.7 ppm increasing with age (temperatures from 0.1° to 8.2°C). In the eggs of pike, *Esox lucius*, it varied from 1.2 to 4.1 ppm depending on age and temperature. This type of finding is of value in assessing the survival chances of fish eggs buried in gravel. *Wicket* (1954), for example, has shown that the flow rates and oxygen content of the water may be quite inadequate for chum salmon eggs, being even as low as 2 mm/hr and 0.2 ppm.

Marine and other fish larvae were studied by *Bishai* (1960a). Herring larvae at hatching were "restricted" at 27-32% saturation and at 55-64% saturation a few days later. Larvae of the lumpsucker, *Cyclopterus lumpus*, just tolerated 43% saturation 3 weeks after hatching and young plaice 4-5 cm in length just tolerated 14% saturation. He found that the larvae of the salmonids, *S. salar* and *S. trutta*, could survive very low levels of saturation, 3-10%, after hatching, but this ability decreased with age, the minimum level tolerated being 16-28%. Again, criteria for defining lethal levels vary, and it is difficult to compare the results of different workers. To some extent acclimation to low oxygen increases resistance to hypoxia and there are also specific differences in tolerance.

Bishai (1960c) has also summarized work on the effects of pH on larval fish. Compared with seawater (pH about 8.0) which is fairly well buffered fresh-water has a pH which is more variable and more susceptible to change by effluents or run-off. Bishai found that herring larvae could survive values between 6.5 and 8.5. Lumpsucker larvae had a low tolerance level of about 6.9, plaice (4-5 cm) 6.2-6.5, while salmonid larvae had a range of tolerance from about 5.8 to 6.2 up to 9.0, or even 10.0 in sea trout alevins. To some extent the levels of tolerance depend on the substances used to change the pH. It is possible that CO_2 for reducing pH has an additional toxic or respiratory effect compared with the use of HCl. Certainly *Alderdice* and *Wickett* (1958) found high levels of CO_2 inhibited oxygen consumption in the eggs of *O. keta*.

Radiation effects on salmonid eggs and larvae are reviewed by *Hamdorf* (1960) and *Eisler* (1961). Strong visible light may cause early hatching mortality, poor growth, and greater pigmentation. Exposure to light is less serious when the intensity is low or in the later stages, especially after the onset of pigmentation. It is, perhaps, not surprising that light is harmful to eggs and larvae which normally develop in complete or semidarkness among gravel or stones; increased activity probably contributes to its deleterious effect. Other species which develop in sand such as the grunion, *Leuresthes tenuis*, or near the sea bed such as herring may also show a poorer rate of hatching under lighted conditions. Careful controls are required in experiments since the light may cause other changes such as raising the temperature or increasing growth of phytoplankton in the water. The main effect on many other species is to accelerate hatching. Ultraviolet light has been tested on *Fundulus* and causes abnormalities of the skeleton, cylopia and twinning. In sockeye salmon, *O. nerka*, it causes premature hatching, vertebral abnormalities, high mortality, and delays pigmentation. *Marinaro* and *Bernard* (1966) exposed the eggs of pelagic fish such as pilchard *Sardina pilchardus*, mullet *Mullus*, horse mackerel *Trachurus* and "sargue" *Diplodus annularis* to sunlight with and without ultraviolet filters, finding a lower rate of hatching with ultraviolet. The transparency of many fish eggs and larvae may be an adaptation to prevent absorption of light as well as acting as a means of camouflage. X-rays have been tested, for example, by Solberg (1938) on *Fundulus* embroys. They were most sensitive after fertilization with a decrease of 10 times in sensitivity between early cleavage and 4-5 days later. Other work is reviewed by Eisler (1961) especially on salmonids. X-rays have a number of effects such as higher mortality, decrease of growth, deformities, decrease in erythrocytes, and destruction of hematopoietic tissue and increased pigmentation. Generally, susceptibility decreases with age and in optimal conditions of temperature. *Neyfakh* (1959) used X-rays for inactivating the nuclei in the developing eggs of the loach, *Misgurnus fossilis*. Irradiation during certain phases of development was followed by death at a very specific time, suggesting a periodical functioning of the nuclei.

Fish eggs in almost any environment are subject to mechanical stimuli which are potentially harmful— movement of gravel within a spawning redd, current borne objects on the sea bed, or wave action at the surface. The resistance of the egg may be measured by the maximum load it will take before the chorion bursts. They indicate an increase in the resistance of the chorion as it hardens subsequent to fertilization and a decrease in the strength of the chorion prior to hatching, presumably resulting from hatching enzymes. Changes in the chorion may in part account for periods of varying resistance to external conditions.

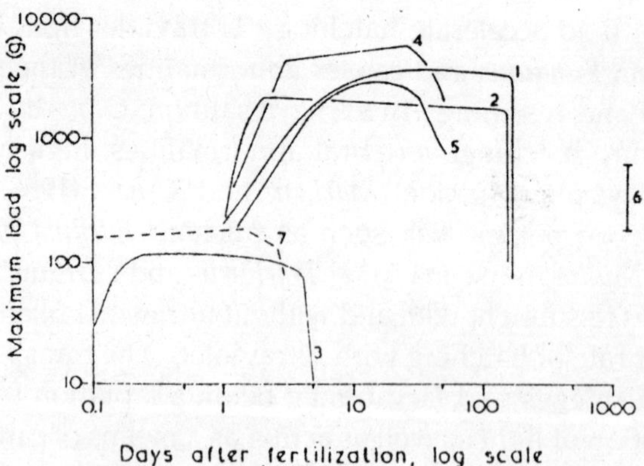

Fig. 4.15.   The maximum load before bursting in eggs of different species during development. 1. *Salmo salar*, 2. *Coregonus lavaretus*, 3. *Acipenser sp.*; 4. *Salvelinus iontinalis* (?); 5. *Salmo salar* ; 6. Pleuronectes platessa (range, varying age) and 7. *Engraulis anchoita*.

Other harmful influences may be hatching enzymes, especially for embryos which may be somewhat unhealthy in other respects, or where eggs are very crowded and the enzymes are concentrated in the water after hatching. *Stuart* (1953) observed sediments attached to the chorion of loch trout. These might well reduce oxygen intake. The alevins, however, seemed to have the ability to disperse such sediments by means of respiratory currents passing over the body surface. With gill respiration sediments were aggregated by mucus and rendered less harmful.

A recent trend in research on tolerance is to use combinations of factors such as temperature, salinity, and oxygen concentration to see how they interact and to find out the optima. *Kinne* and *Kinne* (1962) used such combinations during the development of the desert minnow, *Cyprinodon macularius* (i.e., temperature between 10° and 37°C, salinities from 0 to 85%, and 70%, 100% and 300% saturation of air in water). Mortality was least at near lethal temperatures when the salinity was held between one-half seawater and 35%. The lethal temperature was lower in hypoxial conditions. It seemed that of the salinities used 35% was the optimum for incubation. Lewis (1966), using menhaden larvae, *Brevoortia tyrannus,* found that temperature tolerance was greatest at salinities of 10-15%, suggesting that isotomic conditions were optimum for survival. *Forrester* and *Alderdice* (1966) calculated from experimental data that the optimum salinity and temperature for hatching eggs of cod, *Gadus macrocephalus,* was 19.4% and 5.3°C.

MERISTIC CHARACTERS

Counts of vertebrae, myotomes, scales, gill rakers or fin rays have been used in racial studies. The variability in number and liability under different environmental conditions are striking examples of rather flexible raw material for evolutionary forces. Usually mean counts from samples of fish are required, the differences from race to race being inadequate for identification of individuals. While correlations between the counts in adults or young and the environmental conditions on the spawning ground first suggested that meristic characters were labile, confirmation has been obtained by experimental studies. They show that factors such as temperature, salinity, and oxygen content superimpose their effect on the range of meristic counts determined by the genotype. There is also an individual variation even in fish from the same parents reared under the same conditions. Earlier work, mainly on salmonid and inshore or estuarine species, has been reviewed by Taning (1952) and Barlow (1961). Since it is necessary to keep the young until the meristic characters form there has been less work on species which require to be fed and reared for a considerable time after hatching.

The effects of some factors are given in Table 4.15. Most work has been done on the effect of temperature on vertebral and fin ray counts. Generally, a V-type relationship is found where the average vertebral count is minimum at an intermediate temperature. For example, in sea trout, *Salmo trutta*, Taning found this was at about 6°C; in plaice it is at 8°C; at lower and higher temperatures the average counts were higher. With fin rays the V may tend to be asymmetrically inverted with counts highest at intermediate temperatures. In other instances the mean vertebral counts are inversely related to temperature, but the lack of an inflection may result from an insufficiently wide range of experimental temperatures being used.

Apart from the factors shown in Table 4.13, both intensity and duration of light and CO_2 may have an effect. Higher intensities in the grunion, *Leuresthes tenuis*, reduce the number of vertebrate. A 16-hr day caused lower caudal vertebrae and possibly anal fin ray counts in *Oncorhynchus nerka*, while in sea trout, *Salmo trutta*, increasing tension of CO_2 produced lower numbers of vertebrae.

The work of *Heuts* (1949) and *Lindsey* (1962) using different temperature-salinity combinations in *Gasterosteus aculeatus* were somewhat inconclusive resulting from the problem of getting sufficient numbers of survivors in an extensive series of experiments, some of which were in unfavourable conditions. Certainly, it was shown that temperature effects vary with the salinity level. Using a fresh-water and brackish water race, Heuts found the greatest effect of temperature on fin ray counts at the salinity of adaptation. Differential mortality and 'the unsolved problem of controlling, for example, oxygen and temperature independently have often led to difficulties in interpreting results.

It is the underlying reasons for meristic lability which is, perhaps, of most interest to the physiologist. Sensitive periods have been found during which lability of vertebral counts is as its greatest. In sea trout it is around gastrulation, with a further period when the last vertebrae are being preformed. It is also before hatching in herring and killifish. However, in plaice and paradise fish meristic characters are still susceptible to modification after hatching. Generally speaking, fin ray counts are determined later than those of vertebrae. Later studies on showed that transfer to a higher temperature 2 days after fertilization produced fewer vertebrae, whereas transfer after 6 days resulted in a higher vertebral count. Regular diurnal alterations of temperature between 22° and 30°C gave vertebral counts intermediate between those resulting from sustained temperatures of 22° and 30°C.

Of the various hypotheses put forward to explain meristic lability those of *Gabriel* (1944) and *Barlow* (1961) fit the facts best. Often, but not always, environmental factors which delay hatching produce higher counts. Alternatively, higher counts may be considered to come from nonoptimum conditions. It is likely that the environmental factors change the relationship between growth and differentiation. *Hayes et al.* (1953), in their studies of morphogenesis in salmon at temperatures from 1° to 15°C, certainly found that the relative appearance of certain anatomical characters varied with temperature. If differentiation is late more tissue is available to be differentiated leading to a higher count. Thus one may consider an interaction

TABLE 4.14
Environmental Effects on Meristic Characters by Experiment

Species	Character	Effects of increasing		
		Temperature	Salinity	Oxygen
Lebistes reticulatus (guppy)	Fin rays	Increase		
Salmo trutta (sea trout)	Vertebrae	V relation		
Salmo trutta (sea trout)	Vertebrae	V relation		Decrease
	Fin rays	A-relation(asymmetrical)		
Salmo gairdneri (rainbow trout)	Vertebrae	(?)Decrease		(?)Decrease
	Lateral scales	Decrease		
Oncorhynchus tshawytscha (Chinook salmon)	Vertebrae	V relation		
	Fin rays	A relation		
Fundulus heteroclitus (killifish)	Vertebrae	Decrease		
Macropodus opercularis (paradise fish)	Vertebrae	V relation		
	Fin rays	V relation		
Pleuroncles platessa (plaice)	Vertebrae	Decrease		
Pleuroncles platessa (plaice)	Vertebrae	V relation		
	Fin rays	Increase		
Channa argas (snake-headed fish)	Vertebrae	V relation		
Clupea harengus (berring)	Myotomes	Decrease	Increase	
	Vertebrae	Decrease		
Gasterosteus aculeatus (stickleback)	Vertebrate	V relation	See text	
	Fin rays	Decrease		
Oryzias latipes	Vertebrae	Decrease		

of the Q_{10} for growth and the Q_{10} for differentiation of the character; as the temperature is varied so the relationship between growth and differentiation varies. Where V-shaped curves are obtained there are then points of infection for the temperature coefficients. *Garside* and *Fry* (1959), using normal and reciprocal hybrid fry of speckled trout, *Salvelinus fontinalis*, and lake trout, *S. namaycush*, found that the mean myomere count was lower where the fish developed on the speckled trout yolks, which were smaller. There was also an inverse relationship between myomere count and the degree of twinning in Siamese twins of speckled trout, the count being lower where there was less shared tissue.

REARING AND FARMING

A. Techniques
Three lines of approach are of interest at the present time;

(1)   Tropical and temperate fish culture for food
(2)   Rearing of salmonids and sturgeon to maintain population size after intensive fishing or hydroelectric schemes
(3)   Temperate marine fish rearing and farming.

A good general review on fish farming by Iversen (1968) is available and on tropical species by *Hickling* (1962). Some of the most popular species are the common carp *Cyprinus carpio*, mullets, *Tilapia*, milkfish *Chanos chanos*, the carp *Puntius javanicus*, and grass carp *Ctenopharyngodon idella*. Many of the species are reared in still water fertilized artificially or by sewage. The last four species mentioned are herbivorous which gives maximum efficiency of food utilization. Some of the species, e.g., *Tilapia* and *Chanos*, can be kept in salt or brackish water. In the case of *Tilapia* and the Israeli carp, breeding takes place in the holding ponds; in the European carp special breeding ponds are used. On the other hand, the fry of *Puntius, Chanos, Ctenopharyngodon*, and mullet must be caught in the wild and transferred to rearing ponds. Generally in tropical culture there are no serious problems of feeding the young which are of sufficient size to take the food available.

The extensive breeding of salmonids presents no problem of food supply for the young, which are relatively large when the plentiful yolk has been resorbed. Rearing is usually in running fresh-water although acclimation of the young of *Salmo* species to marine conditions is now being developed, with improved growth and freedom from disease. The question of dietetics is well advanced in the group.

Marine fish rearingfarming, although attempted from time to time over the last hundred years, have only recently made significant advances. Starting with the liberation of millions of plaice and cod eggs or yolk sac larvae in coastal areas, only the later experiments by *Dannevig* (1963) seem convincingly exploitable. Very recently, *Mugil* and *Solea* have been established as self-perpetuating species in some of the Mediterranean coastal lagoons.

Following the earlier rearing in aquaria of some species to an advanced stage by highly empirical methods, a better experimental approach has now been adopted. Success in rearing *Pleuronectes platessa, Solea solea, Microstomus kitt Clupea harengus Sardinops caerulea, Engraulis* and *Scomber* and *Sardina pilchardus* has depended on some or all of the following factors:

(1)   *Food supply.* The use of *Artemia* and *Balanus* nauplii, *Mytilus* trochophores, young *Tigriopus*, small nematodes, rotifers and oligochaetes and secondary foods like Dunaliella has improved survival rates, so also has the extensive collection and selection of natural plankton.

(2)   *Antibiotics.* Particularly with flatfish, dosages of mixed sodium penicillin (50 IU/ml) and streptomycin sulfate (0.05 mg/ml) have improved survival in the egg stage.

(3)   Black-walled tanks and uniform overhead illumination have helped feeding success by providing contrast of the food against the background and even distribution of food in the tank.

(4)   Tank hygiene.

There are interesting physiological and behavioural aspects of rearing where the fish may be artificially crowded and conditions of stress set up. Size hierarchy effects prevail but growth seems to be density dependent only from the beginning of metamorphosis. Bitten fins are found more often in the smaller members of a tank community but are not always dependent on density. Perhaps more serious is the incidence of abnormal (albino or semi-albino) pigmentation in flatfish. It is generally worse in smaller fish and in very crowded tanks but may be reduced by using *spawning* stock acclimated to hatchery conditions. Survival to metamorphosis is also better in larvae from acclimated spawning stock.

B. Sensory Deprivation

Hatchery-reared flatfish do not survive well when transferred to the sea, presumably because of high predation; hatchery-reared salmonids usually have a poorer swimming performance. In fact, the hatchery environment by preventing contact from predation, by often rather uniform physical and chemical conditions and lack of shelter may prevent the development of appropriate defensive responses and muscular systems. The need for high density may also encourage undesirable intraspecific relationships at the expense of the more desirable interspecific responses.

The more obvious shortcomings of the aquarium tank may be accompanied by sensory deprivation in a more general and less easily defined way. The uniformity of the aquarium environment may lead to insufficient sensory input. For instance, *Qasim* (1959), when rearing the young of *Blennius pholis*, obtained better survival with a day-night regime than with continuous light. *Blaxter* (1968a) rearing herring, found a hint of better survival when the light conditions over the tank were continually changed and when air jets playing on the surface kept up a continuous series of ripples. *Hoar* (1942) found that young salmon would feed on chopped earthworm in darkness during the day, but not at night, perhaps because of an activity (or retinomotor) rhythm. Such rhythms might eventually be suppressed by continuous light. Activity rhythms with periods of rest might play as important a part in the development of the sense organs and nervous system as sufficient sensory input during the active phases.

5

Yolk

A fish egg can be considered a semiclosed system. Once the egg membrane(s) has been hardened by exposure to water, the membrane(s) permits egg exchange but is relatively impervious to most solutes. As a consequence, the majority of fish embryos are dependent on endogenous yolk reserves to supply the substrates for energy production and growth. Viviparous fishes are an exception. Both the rate of yolk absorption and the efficiency of yolk utilization are important determinants of early development, growth and survival. Larval survival is ultimately dependent on the availability of food in sufficient quantity and of adequate quality after yolk reserves are exhausted. It follows that there are strong selective pressures synchronizing completion of yolk absorption, development of the capability of feeding, and the availability of suitable food. Since large size confers certain advantages on larvae, there are strong selective pressures to maximize the efficiency with which yolk is converted into tissues. Larger larvae of a given species can be expected to be stronger swimmers less affected by competition more resistant to starvation less susceptible to predation able to commence feeding earlier and able to have increased success at first feeding.

A number of environmental factors like temperature, light, oxygen concentration, and salinity influencing the rate and efficiency of yolk absorption. Fish eggs are not motile, however, and thus developing embryos are unable to actively exploit the most favourable environments available, at least until after hatching.   Only species that utilize reproductive strategies such as viviparity or mouth brooding may, through parental behaviours, be able to manipulate *egg* incubation conditions. It is selectively advantageous, therefore, for a species to produce eggs that can develop successfully within a range of *"expected"* incubation conditions. The scope of these *"expected"* conditions will depend on those conditions experienced during evolution of the species. For some fishes, the fluctuation in environmental parameters may be relatively slight (e.g., abyssal marine habitats), while for others it may be large (e.g., some temperature fresh-water habitats).

STRUCTURAL ASPECTS OF YOLK ABSORPTION

A. Yolk Morphology

The fish yolk includes yolk platelets and oil globules as the structural components. The majority of yolk platelets are round or oval in shape, flattened in one plane, and 4-15 μm in length. Larger platelets appear to be characteristic of species possessing larger eggs. The size of platelet, also varies within each egg, with the deeper, more centrally located platelets tending to be larger and more homogenous than the superficial peripheral ones. Each platelet consists of an outer sheath and a central core. The sheath forms a semipermeable bilayer around the core and contains mucopolysaccharides. The core is composed of lipovitellin and phosvitin, or analogous lipoproteins and phosphoproteins. These core proteins may or may not be arranged in a crystalline lattice. Moreover, the crystalline structure may be lost as the ova mature. Oil globules are located among the yolk platelets. Globule number and size vary greatly among species, from innumerable small globules in the micrometer diameter range to singular large globules in the millimeter diameter range. The globules contain primarily triglycerides, although proteins wax esters and carotenoid pigments are also present in some species.

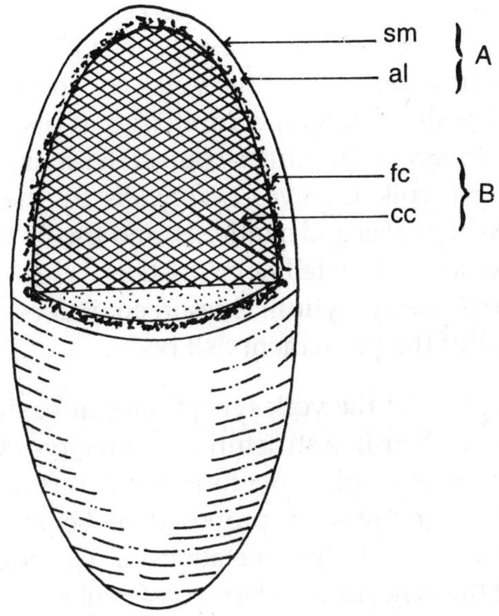

Fig. 5.1.   Model of a yolk platelet based mainly on the structure of *Amia* platelets. The cuts reveal its interior: (A) outer sheath; (B) main body; sm, superficial membrane; al., amorphous superficial layer; fc, fibrillar cortex; cc, crystalline core.

B. Meroblastic Fishes

In most of elasmobranchs and teleosts the meroblastic cleavage, results in the formation of an extraembryonic yolk sac. A characteristic feature of this extraembryonic sac is the *yolk syncytium*, a specialized tissue responsible for absorption of yolk. The presumptive yolk syncytium, the *periblast*, is recognizable in the fertilized teleost egg at the one-cell stage. As cleavage proceeds, numerous free nuclei appear in the periblast, thus transforming the layer into a true syncytium.

In teleost eggs, the *yolk syncytium* together with overlaying mesoderm and ectoderm spreads to enclose the entire yolk mass. Endoderm does not follow the movement of the teleost blastodisc rim and, consequently, the yolk is not enclosed by an endodermal layer. Absorption of yolk nutrients in teleosts, therefore, occurs without any involvement of endodermal cells or the gut. A system of blood vessels, the vitelline circulation, develops within the walls of the yolk sac. In some areas, the endothelial wall of vitelline capillaries is incomplete and embryonic blood is in direct contact with the syncytium. Absorption of yolk involves *endocytosis* by the syncytium, *intrasyncytial digestion* and synthesis, and finally the release of yolk metabolites to the vitelline circulation. When yolk reserves are exhausted, the syncytium is resorbed; it does not take part in formation of the permanent fish body.

Two regions of the yolk syncytium can be distinguished on the basis of their fine structure. One region, characterized by smooth endoplasmic reticulum, numerous mitochondria, and glycogen granules, is proposed to be responsible for carbohydrate and / or lipid metabolism. This region extends throughout the syncytium. The second region is characterized by rough endoplasmic reticulum and Golgi complexes, and extends in portions across the syncytium forming a stratified structure. This latter region is thought to be involved in the synthesis and transport of proteinaceous substances. Yolk protein must be dephosphorylated to become soluble. *Amirante* (1972) suggested that fish yolk proteins are solubilized by the action of calcium and phosphoprotein phophatase. Syncytial Golgi complexes probably supply acid hydrolases for the degradation of yolk platelets.

In addition to the syncytial layer, the yolk itself contains enzymes that probably facilitate the breakdown of yolk into its constituent nutrients. *Vernier* and *Sire* (1977) described two types of yolk platelets with different enzyme contents. One form, the embryonic platelet type, has an enzyme load that allows nutrients to be released prior to establishment of the syncytium. The second or usual platelet type lacks this enzyme load and is digested by syncytial enzymes.

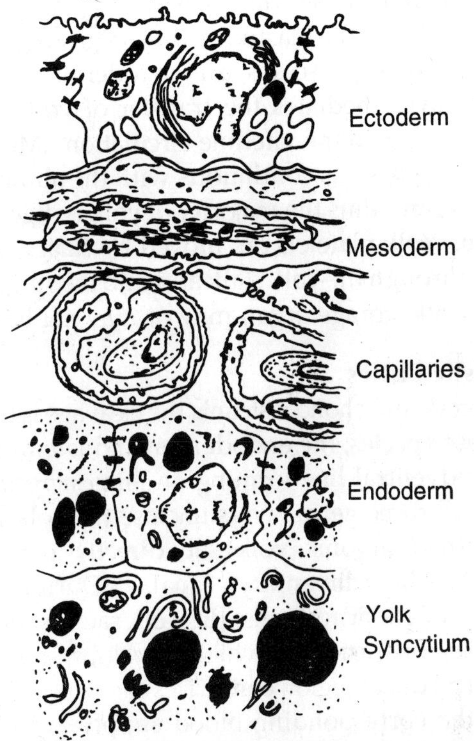

Fig. 5.2. Idealized diagram of the cellular organization in a preimplantation shark yolk sac. The teleost yolk sac is similar in structure except that it lacks endoderm.

In teleosts the extraembryonic yolk sac with its yolk syncytium is the sole site of yolk absorption but this is not the case in chondrichthyean fishes including sharks, skates, rays and ratfish. In holocephalians, for instance, only a small portion of the yolk mass is enclosed by the yolk sac. The remainder breaks up into a viscous fluid, which is first absorbed via the external gills of the embryo and later ingested through the mouth. This ingestion of yolk nutrients is comparable, in a general sense, to that exhibited by *oophagous sharks*, whose viviparous embryos ingest ova present in the same uteri.

In cartilagenous fishes the formation of an archenteron at the posterior edge of the blastodisc during gastrulation results in a *yolk sac* that possesses an endodermal layer. This endodermal layer mediates the transfer of yolk metabolites from the *syncytium* to the vitelline circulation. Moreover, the elasmobranch *yolk sac* is continuous with the alimentary tract via a *yolk stalk*, and thus the majority of yolk is digested within the intestine. Yolk platelets are moved by ciliary action from the yolk sac through the yolk stalk and into the spiral intestine. An internal yolk storage organ may or may not be present.

C. Holoblastic Fishes

In lampreys and chondrosteanis the cleavage is of *holoblastic* type. In these species, the endodermal and lateral plates fuse along the midventral line forming an *intraembryonic yolk sac*. As a result, all three germ layers enclose the yolk mass. The resultant intraembryonic yolk sac directly participates in formation of the alimentary canal. During posthatch development of chondrosteans, the yolk sac is separated into two major regions, each of which develops separate blood drainages. The distal region comprises the intestine and spiral valve, and the corresponding blood supply proceeds to the liver. Yolk within this region is the first to be utilized. The second region includes the stomach and esophagus and develops a blood supply that proceeds directly to the sinus venosus. This region is the last portion of the alimentary canal to differentiate, and yolk is retained there longer.

Although *hydrolytic enzymes* are present within the developing alimentary canal, their activities are low. The existence of yolk material within endodermal cells lining the yolk sac implies that *endocytosis* and *intracellular digestion* may be the primary mechanisms by which yolk nutrients are made available. There is a similar mechanisms for mobilization of yolk nutrients in the intraembryonic yolk sac of holoblastic fishes and the extraembryonic yolk sac of meroblastic fishes.

YOLK COMPOSITION DURING DEVELOPMENT

In fishes there is a wide diversity or reproductive strategies. As a consequence, egg size and fecundity vary among species. In *oviparous fishes* from about 0.7 mm egg diameter (e.g., convict surgeon fish *Acanthurus triostegus*) to greater than 10 mm diameter (e.g., chinook salmon *Oncorhynchus tshawytscha)*, with spawns varying from less than 100 eggs per female (e.g., mouthbrooding cichlid *Labeotropheus fuelleborni)* to more than 9,000,000 (e.g., Atlantic cod *Gadus morhua*). Viviparous fishes tend to produce fewer but proportionally larger eggs. These differences in egg size and number imply maternal investment per egg differs widely among species. The nutrient composition of fish eggs is species-specific. Within a given species, as well, egg quality varies as a function of maternal egg, weight and diet. Despite these differences, the dynamics of yolk absorption are similar among groups. Following fertilization, the developing embryo begins to utilize yolk nutrients. This is accompanied by increasing consumption of oxygen, particularly after the blastula stage is reached. As development proceeds, the absolute and relative composition of the yolk changes.

Various approaches (proximal analysis, respiratory quotients, radiolabeled substrates) have been used to investigate the sequence with which yolk nutrients are catabolized for energy production. Generally, carbohydrate, lipid, and protein are consumed prior to hatching, while lipid and protein catabolism predominate after hatching. However, the precise sequence of nutrient consumption varies both qualitatively and quantitatively. This is not surprising, since it is unlikely that any one energetic scheme is adequate to describe the sequential utilization of energy substrates in all fishes, in other than general terms. More likely, the precise scheme varies among fishes in relation to absolute egg composition.

TABLE 5.1
Chemical Composition of Fish Eggs

Species	Dry Weight		Percentage of dry weight			
	mg	%	Protein	Lipid	Carbohydrate	Ash
Acipenser transmontanus (white sturgeon)	6.25[a]	23.8	67	30	--	3
Coregonus albula (vendace)	16.27[b]	—	64.4	25.8	—	8.5
Coregonus lavaretus (whitefish)	15.6[a]	—	60.3	27.7	—	9.8
Cyprinus carpio (carp)	—	30.4[a]	64.3	5.9	3.7	--
	—	10.2[b]	58.3-59.2	5.4-29.3	1.5-6.2	6.3
Eleginus navaga (navaga)	0.283[a]	22.1	66.4	20.5	—	2.1
	0.298[b]	10.4	56.7	16.8	—	8.4
Morone saxatilis (striped bass)	0.232[a]	46.3	—	52.0	—	3.0
Pseudopleurconectes americanus (winter flounder)	0.51AF[a]	—	79.3	15.4	5.3	--
	0.049AF[b]	—	77.4	19.4	3.2	7.2
Salmo gairdneri (rainbow trout)	42.1[a]	41.3	56.2	—	—	--
	—	33.8[b]	59.8-71.3	11.4	0.6	3.8-3.9
Salmo salar (Atlantic salmon)	49.7[b]	36.0	52.2	36.1	1.0	2.8
Sardinops caerulea (Pacific sardine)	—	29.3[a]	71.6	13.0	<1	7
Sciaenops ocellata (red drum)	—	7.0[b]	28.1	33.7	0.4	--

[a] Unfertilized.
[b] Fertilized, AF, ash-free.

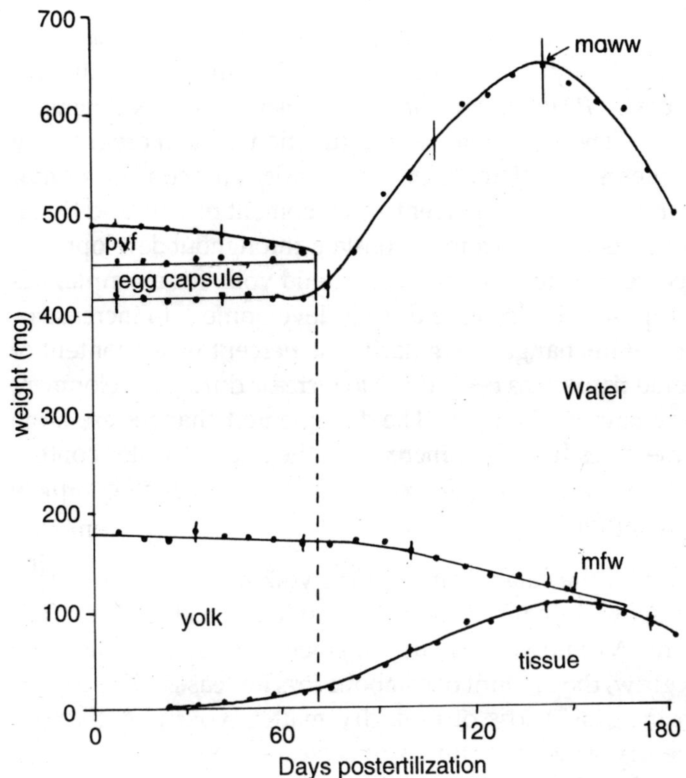

Fig. 5.3.   Typical changes in the relative composition of fish during the yolk absorption period. Data are from chinook salmon (*Oncorhynchus tshawytscha*) at 8°C. The broken line represents 50% hatching. Variability about some means is indicated by 95% confidence limits; pvf, perivitelline fluid; maww, maximum eleutheorembryo wet weight; mtw, maximum tissue weight.

A. Dry Matter and Water Content

The percentage dry weight per egg, and hence egg water content, is variable. Following fertilization, egg water content comes under osmoregulatory control. The specific gravity of fish eggs is inversely related to water content. Thus, the buoyant nature of pelagic eggs is associated with a higher water content relative to that of dense, less hydrated demersal eggs. The specific gravity of eggs and eleutheroembryos can decrease with time [e.g., Atlantic herring *Clupea harengus harengus*, Atlantic salmon *Salmo salar*,] or increase with time [e.g., plaice *Pleuronectes platessa*, northern anchovy *Engraulis mordax*]. These changes are a function of osmoregulatory adjustments and changes in the quantity and composition of yolk and tissues. The percent water content of yolk and tissue cannot be assumed to remain constant-throughout development. The percent water content of salmonid yolk, for example, has been reported to decrease during development to increase or to remain unchanged. Similarly, the percent water content of salmonid tissues has been found to increase during development or to remain unchanged. The documented changes are large in some cases [e.g., 20% increase in the relative water content of rainbow trout *(S. gairdneri)* yolk from hatch to complete yolk absorption.

Due to the catabolism of some yolk materials for energy production, the conversion of yolk to tissues is less than 100% efficient. As embryonic tissues and their associated maintenance costs grow, the amount of catabolic loss increases. This results in an increasing loss of bulk dry matter (yolk plus tissues). Tissue dry weight continues to increase, however, as yolk is absorbed and converted into tissues. As yolk reserves near exhaustion, the metabolic demands of maintenance and activity exceed the supply of yolk nutrients, and tissues begin to be catabolized for energy production. The resultant reduction in body weight causes a maximum tissue weight to be reached before yolk absorption is completed. *Wallace* and *Aasjord* (1984b) found that Arctic char *(Salvelinus alpinus)* at 3°C reached maximum tissue weight at complete yolk absorption, whereas at 12°C maximum tissue weight was reached with 1.5 mg of dry yolk remaining (12% of yolk reserves at hatching). The timing of maximum tissue weight, and hence the end of growth utilizing yolk alone, would appear to be influenced by temperature, occurring earlier in development at higher temperatures. This temperature dependency may reflect changes in maintenance costs or shifts in the relative rates of protein and lipid mobilization from yolk. The temperature effect may also explain why a maximum tissue weight has not been evident prior to complete yolk absorption in some studies.

The bulk wet weight of the embryo/eleutheroembryo increases during development despite the concurrent loss in bulk dry weight, because of uptake of water. *Gray* (1928) modeled these early weight changes and predicted that, dependent on metabolic costs, the bulk wet weight would reach a maximum before the tissue growth cycle was completed. Thus, toward the end of the endogenous nutrition period, eleutheroembryo wet weight can be expected to decrease despite a continued increase in tissue weight; the resultant maximum eleutheroembryo wet weight will be reached before the maximum tissue weight. Maximum eleutheroembryo wet weight is reached earlier in development when relative metabolic costs are elevated, such as at higher rearing temperatures or in smaller eggs of a given species.

B. Protein

Protein is the most abundant dry constitutent of many fish eggs. Yolk protein serves two primary functions: it provides amino acids for tissue growth and supplies energy via catabolic processes. As a result, there is a continual loss of protein from the yolk mass as it is transferred to the developing tissues. In addition, the catabolism of protein for energy production results in a decline of bulk protein quantities (tissue plus yolk). During early development, when the embryo is small and total metabolic activity is low, little of any protein is utilized for energy production; bulk protein quantities remain relatively constant. As growth proceeds and the metabolic rate increases, a larger portion of yolk protein is shunted into energy production and bulk protein quantities decline. This is especially evident after hatch, in accordance with the higher levels of activity and energy demand. Studies of oxygen consumption and nitrogen metabolism also support an increased use of protein for energy production after hatching. It is difficult, however, to directly determine protein catabolism before hatching because of the low permeability of egg membranes to nitrogenous metabolites.

TABLE 5.2

Relative Distribution of the Constitutents of Freshly Fertilized and Water-Hardened Eggs of Atlantic Salmon (*Salmo salar*)

		Percentage of total weight		
Constitutent	*Total weight (mg)*	*Egg membranes*	*Perivitelline fluid*	*Yolk*
Water	88.3	5.2	23.3	71.8
Dry matter	49.7	0.8	16.9	82.1
Protein	26.0	0.3	22.7	76.9
Lipid	18.0	0.1	7.7	92.2
Carbohydrate	0.5	48.0	46.0	2.0
Ash	1.4	7.1	7.1	85.7

There are three major classes of yolk proteins. In order of abundance, these are *lipoproteins*, *glycoproteins* and *phosphoproteins*. During development, there appears to be little change in the relative proportion of these yolk proteins. This suggests the three protein types are utilized at the same relative rate. However, the physical and chemical properties of yolk lipoprotein change during development; molecular weight increases, the prosthetic lipid groups decrease, and specific amino acids are released. Fish embryos develop at the expense of yolk protein degradation and the liberation of amino acids. Changes in amino acid concentrations during development appear to parallel changes in the concentrations of DNA and RNA.

C. Lipid

Following protein lipids are the next most abundant dry constitutent of most fish eggs. There is considerable interspecific variation, however; *Balon* (1977) lists lipid quantities ranging from 0.1% of egg weight in the plaice *P. platessa* to 45% in the mouth-brooding cichlid *Labeotropheus*. This variation has been considered an adaptation to different reproductive strategies or to the duration of the endogenous nutrition period. Lipid content and the presence of large oil globules are poorly correlated with the pelagic or demersal nature of eggs, and thus lipids probably do not function primarily in buoyancy. During development, lipids are converted into structural components such as cell membranes or channeled into energy production. The amount of catabolic loss increases as the fish body grows, but is relatively insignificant until hatch and thereafter. This is probably related to increased activity and the resultant higher energy requirements following hatch. On a caloric basis, lipids, especially triglycerides and wax esters (neutral lipids), are the most important energy reserve of developing fish.

Nakagawa and *Tsuchiya* (1971, 1972) describe two major lipid types : (1) *free lipids* associated with the oil globule, and (2) *bound lipids* associated with the high-density fraction (HDF) and therefore primarily within the yolk platelets. The oil globules are composed primarily of *triglycerides; phospholipids* are not detectable in salmonid globules but may be present in the oil globule of thermophilic fishes. In contrast, lipids of the HDF (yolk platelets) are dominated by phospholipids with triglycerides being secondary in abundance. It is interesting to note that wax esters are a major lipid component of the yolk of some species. In red drum eggs *(Sciaenops ocellata)*, for example, wax esters account for 29% of the total lipid pool and provide 53% of the total calories consumed between fertilization and hatching. *Vetter et al.* (1983) noted that species incorporating large quantities of wax esters into their ova produce buoyant eggs that commonly encounter reduced salinities. They hypothesized that, since wax esters have a lower specific gravity than triglycerides, these esters play an role in egg buoyancy.

The yolk of many species contains lipid-soluble carotenoid pigments. These pigments are concentrated in the oil globules but are also present to a lesser extent in the HDF (yolk platelets). Yolk carotenoids may represent a nutrient and/or may function for protection from sunlight and for respiration.

The majority of studies concerning lipids have examined bulk changes within the whole embryonic system and have not distinguished between yolk and tissues. It is generally difficult, therefore, to determine if there is any selection, retention, or utilization of specific yolk lipids (triglyceride, phopholipid, wax esters), and whether interconversion between yolk lipids is possible. Data from bulk studies reveal that there is little change in relative lipid composition during embryonic development. This indicates that prior to hatch the major lipid classes are utilized at the same relative rate with little or no preference for specific lipids. Following hatch, the decline in bulk lipid quantities is mainly due to catabolism of triglycerides, while phospholipids, which would be incorporated into structural components (membranes) of the developing fish are conserved. Analyses of yolk separate from tissues and egg membranes, however, indicate that after hatch the lipids of the yolk platelets (phospholipids) are preferentially consumed over these of the oil globule (triglycerides).

Shifts in fatty acid composition of the lipid classes also occur following hatch. This demonstrate the existence of preferential retention and utilization of certain fatty acids and the possibility of some fatty and synthesis by eleutheroembryos. These trends are evident in bulk studies, but are made more lucid when yolk is analyzed separately from the tissues. For example, essential fatty acids of the linolenic series and other polyunsaturated fatty acids that cannot be metabolically synthesized by fish are concentrated within the tissue to levels higher than those in the yolk. In contrast, the concentrations of some fatty acids are higher in the yolk than in the tissues. As a result, fatty acid composition of the tissues is not directly related to amounts within the ova or within the yolk at later developmental stages.

Some of the changes in fatty acid composition, particularly at hatch, reflect the establishment of fatty acid synthetase systems. This late onset of synthetic activity may result in embryos having a more extensive fatty acid requirement than juvenile or adult fish. Gourami embryos, for example, apparently require both linolenic and linolenic acids, whereas only linolenic is essential to older fish. De novo synthesis of fatty acids by the embryo, however, may not be energetically efficient.

D. Carbohydrates

Fish eggs contain relatively little carbohydrate when compared with the amounts of protein and lipid. Moreover, a large proportion of total egg carbohydrate is associated with the egg membranes and therefore, is probably unavailable for use by the developing fish, at least until hatching. Yolk carbohydrates are present in both a free state and complexed with yolk proteins. In rainbow trout (*Salmo gairdneri*) yolk, glycoproteins represent more than 10% of the total protein pool. Glycogen is the primary carbohydrate in all fish eggs studied to date.

Carbohydrate tends to be the most utilized yolk nutrient. In red drum (*Sciaenops ocellata*), for example, the glycogen pool decreased 63% between fertilization and hatching yet because of the small size of the pool accounted for only 1.7% of the total calories consumed. Intense catabolism of carbohydrate commences at fertilization indicating that carbohydrate plays an important nutritive role during initial cleavage. Its importance to later developmental stages has probably been underestimated by past bulk studies because of the early establishment of gluconeogensis. Glycogen, for example, is accumulated in embryonic liver cells in significant amounts prior to hatching. Such carbohydrate reserves might be expected to be important energy sources during periods of tissue hypoxia when aerobic lipid metabolism is not possible.

E. Caloric Content

The chemical composition of yolk changes during the

course of development and, hence, its caloric content can be expected to change. *Kamler* and *Kato* (1983) recorded a decrease in the energy content (calories per milligram dry weight) of rainbow trout *(Salmo gairdneri)* yolk from 6.675 at fertilization to 6.344 at hatching. This suggests a decrease in the relative proportion of yolk lipid to protein prior to hatch. On a dry-weight basis, oil globules contain approximately 1.7 times the energy of yolk platelets, in accordance with the predominance of lipid in the former and protein in the latter. Yolk platelets are mobilized more rapidly than the oil globule from the yolk mass, especially after hatching. Consequently, after hatching, the relative caloric content of the yolk mass (platelets plus

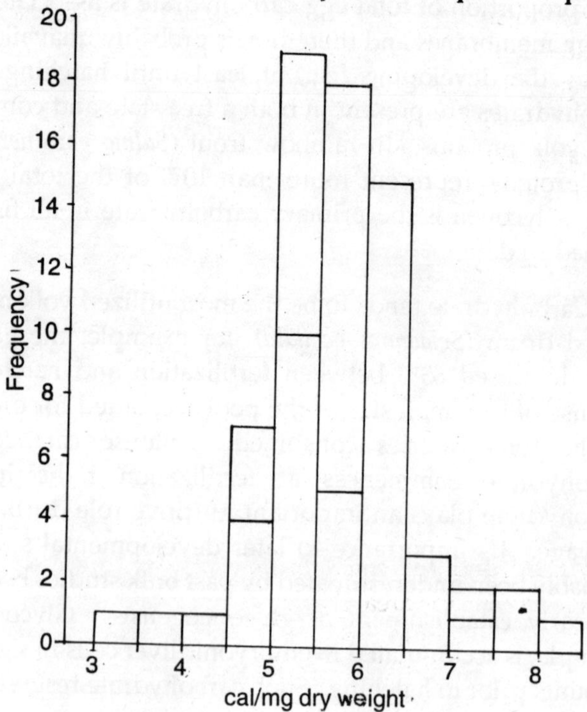

Fig. 5.4   Frequency distribution of the caloric content of eggs or ripe ovaries of fish, based on 70 observations from 54 species.

globules) can be expected to increase. This has yet to be directly ascertained. Nonetheless, the available data do not during development. This argues against the use of a single caloric content value (calories per milligram yolk) in energetic calculations of the rate and efficiency of yolk absorption.

ABSORPTION OF YOLK

The rate of yolk absorption can be determined by following the change with time in yolk calories, dry weight, wet weight, volume, or planar area. Each of these methods is valid within certain limiting conditions. Volume and area determinations are generally made from only two measured dimensions, which introduces unknown error. These techniques also require that the entire yolk mass be visible and, thus, have limited application for heavily pigmented eleutheroembryos. As well, yolk volume, area, and wet weight can be influenced by environmental factors and specimen preservation. Yolk wet weight will be influenced by the relative water content of yolk, which may change during development. From an energetic point of view, measurement of total yolk calories is the best approach. However, since yolk composition and caloric content per unit weight change during development, this approach requires separate determinations of yolk caloric content at each sampling time.

The rate at which yolk reserves are depleted must be a function of the surface area of the absorptive layer (e.g., yolk syncytium) and the metabolic activity of that layer. The absorptive surface area changes during development, being minute at fertilization and then expanding to enclose the entire yolk mass in most fishes. In teleosts, the absorptive surface area is approximately equal to the area of the yolk sac. Hence, as yolk reserves are depleted and the yolk sac decreases in size, the absorptive surface area must also diminish. The reduction in surface area can be temporarily ameliorated by concurrent changes in yolk mass shape. Thus, at any given time, the surface area available for yolk absorption is dependent on the size and, to a lesser degree, shape of the yolk mass.

The surface area and volume of asymmetric yolk masses, such as that of the white sucker *(catostomus commersoni)*, which has a bulbous anterior segment and an elongated posterior segment, can be estimated by addition of the appropriate formulas, in the case of white sucker the formulas for a sphere and a cylinder. In nonteleost fishes that utilize the alimentary tract for yolk absorption, the surface area available for yolk absorption is independent of yolk sac size or shape.

The relationship between yolk volume and absorptive surface area explains, in part, why the absorption rate of a given species is more rapid in larger eggs than in smaller eggs. In salmonids, the rate of yolk absorption appears to vary in a 1 : 1 relationship with egg size. Thus, all other factors being equal, eggs of a given salmonid species reach complete yolk absorption within a span of several days, despite large differences in egg size. This may be a unique adaptation related to the reproductive strategy of salmonids. In cod and herring, the relationship between relative absorption rate and relative egg size is closer to 1 : 2. Thus, a doubling of egg size in these latter species prolongs the period of endogenous nutrition (fertilization to complete yolk absorption) by about 1.3 times. In these latter species, the rate of yolk absorption per unit area of syncytium must decrease as egg size increases.

The teleosts exhibit three distinct phases of yolk absorption. The first or prehatch phase is characterized by slow but steadily increasing rates of yolk absorption. Yolk platelets and oil globules are consumed at approximately the same relative rate during this phase. Shortly before and at hatching, the rate of yolk absorption increases rapidly, probably in response to both an increase in absorptive surface area due to changes in yolk sac shape and an increase in the metabolic activity of the yolk syncytium. This marks the beginning of the second or posthatch phase of absorption, which is characterized by a relatively high and constant rate of absorption. During the posthatch phase, yolk platelets are preferentially consumed over the oil globule.

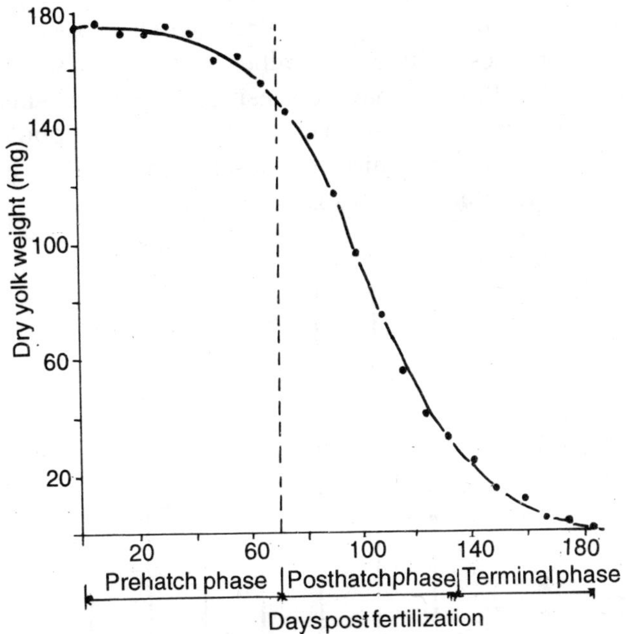

Fig. 5.5. Typical changes in dry yolk weight of teleost fish. Data from chinook salmon (Oncorhynchus tshawytscha) at 8°C. The period of endogenous nutrition (fertilization to complete yolk absorption) has been divided into three phases based on trends in the rate of yolk absorption. The broken line represents 50% hatching.

The rate of yolk absorption slows as the reserve of yolk platelets nears exhaustion probably in response to both a decrease in absorptive surface area as the yolk sac shrinks and the changing composition of yolk. This marks the beginning of the terminal phase of absorption, during which the remaining yolk, predominantly oil globules, is consumed.

Factors that increase or decrease the metabolic activity of

the yolk syncytium can be expected to increase or decrease, respectively, the rate of yolk absorption. The rate of yolk absorption is reduced, for example, by low dissolved oxygen concentrations, sub-and supraoptimal salinities, high ammonia concentrations, and sublethal concentrations of toxic xenobiotics. Some xenobiotics induce deformities in the yolk sac. The structure of yolk itself may be sensitive to some chemicals; fuel-oil fractions can cause coalescence of the oil globules in fish yolk. The extent to which yolk absorption is influenced by such structural abnormalities is unclear.

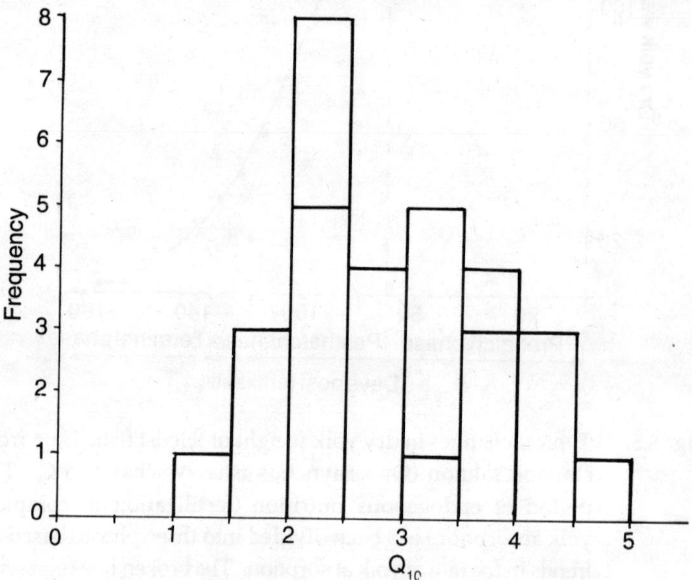

Fig. 5.6. Frequency distribution of the Q^{10} values for yolk absorption in fish, based on 29 observations from 23 species. Shaded areas designate marine fish eggs. (From numerous sources).

Temperature has a differential effect on the absorption of yolk platelets and oil globules. Oil absorption appears to be affected more than platelet consumption by increases in temperature. Near the lower limit of the tolerated thermal range, early life stages may encounter problems with oil absorption and metabolism, and platelet consumption may

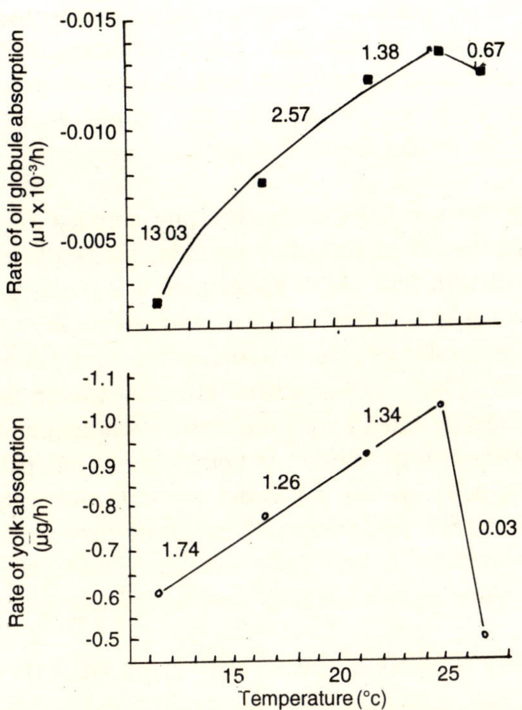

Fig. 5.7. Temperature-specific yolk (=yolk platelets) and oil globule absorption rates in California grunion (*Leuresthes tenuis*). The Q10 values for each incremental increase in temperature are shown in the figure.

dominate. In some species, the rate of yolk absorption is also sensitive to mixed feeding (exogenous plus endogenous). In striped bass (*Morone saxatilis*), for example, mixed feeding results in increased consumption of the oil globule starved larvae conserve their oil globules, yet still experience tissue resorption and eventually die with oil remaining. The effect of mixed feeding on oil consumption probably reflects an increased catabolism of oil reserves to meet the energetic demands associated with feeding activity. On the other hand, feeding activity has no effect on the yolk absorption rate of chinook salmon (*Oncorhynchus tshawytscha*). In still other studies, mixed feeding has been shown to slow the rate of yolk absorption [e.g., walleye pollock *Theragra chalcogramma*, Arctic cod *Boreogadus saida*. In this latter group of fishes, utilization of exogenous nutrients not only satisfies the metabolic costs associated with feeding activity but would also appear to influence the utilization of endogenous nutrients.

It has been noted that swimming activity dramatically accelerated the rate of yolk absorption in ayu eleutheroembryos (*Plecoglossus altivelis*). After swimming at a cruising speed of 0.3 body lengths s^{-1} for 60 min, the mean yolk volume of active ayu eleutheroembryos was 36% smaller than that of unexercised control fish. This is not a general effect, however, because in other species the rate of yolk absorption is independent of the energetic demands of activity. In *Oncorhynchus* spp. particularly, swimming activity has no effect on yolk absorption rate. Moreover, *Hansen* and *Moller* (1985) found the opposite effect in Atlantic salmon (*S. salar*); active eleutheroembryos absorbed their yolk reserves more slowly than inactive eleutherembryos.

EFFICIENCY OF YOLK UTILIZATION

The efficiency of yolk utilization is measured in terms of the growth sustained by yolk absorption. Efficiency is commonly calculated as the ratio between the change in tissue dry weight or calories and the concurrent change in yolk weight or calories. In addition to growth, absorbed yolk supports differentiation, maintenance and activity. Yolk utilization is less than 100% efficient, therefore, due primarily to the metabolic costs of maintenance and activity. The costs of differentiation are probably constant among individuals of a given species and relatively small, and hence can be ignored.

The rate and pattern of embryonic growth are functions of the following : yolk composition, yolk digestion by the syncytium or analogous tissue; the uptake and transport of yolk nutrients from the yolk mass to the developing tissues: activity of the somatic synthetic machinery; and the metabolic demands of maintenance and activity. Factors acting at the level of the yolk sac can be expected to manifest themselves as changes in yolk absorption rate but need not influence utilization efficiency. In other words, yolk absorption is slower, but the fish size ultimately attained is unchanged. On the other hand, factors acting at the level of the somatic tissues can be expected to manifest themselves as changes in utilization efficiency but need not influence absorption rate, that is, the timing of yolk exhaustion is unchanged, but the ultimate fish size is reduced. Generally, early life stages utilize their yolk reserves more efficiently than later life stages utilize exogenous food. The available data indicate that, under optimal conditions, yolk utilization efficiencies can be as high as 60-90% for both dry-matter conversion and caloric conversion.

A. Biotic Factors

The metabolic demands of maintenance and activity vary during development. Maintenance costs increase as growth proceeds. The costs associated with activity are less predictable, varying among individuals depending on the level of spontaneous activity. It follows that the efficiency of yolk utilization is not constant throughout development. Efficiency, in fact, reaches zero and then becomes negative as the maximum tissue weight is reached and tissues are then resorbed during the terminal phase of yolk absorption. For this reason, comparison of gross efficiencies calculated from differing segments of development (e.g., fertilization to hatching versus hatching to complete yolk absorption) are of questionable validity. There is little agreement, however, as to exactly what constitutes an equivalent segment of development. Hatching is not a developmental event *per se* and therefore should be used with caution when calculating efficiencies. Effects of temperature on the developmental timing of maximum tissue weight and maximum eleutheroembryo wet weight complicate matters still further. If one assumes that larval size at first feeding is an important determinant of subsequent growth and survival, perhaps the most relevant determination is the gross efficiency between fertilization and the time of 50% feeding, independent of the developmental stage at which 50% feeding occurs.

Yolk utilization is influenced by egg quality and yolk composition. *Rogers* and *Westin* (1981) found, for example, that unfed striped bass *(Morone saxatilis)* conserved their oil reserves, yet still experienced a metabolic deficit during the terminal phase of yolk absorption. The data suggest that tissue resorption during the terminal phase of yolk absorption was due to preferential depletion of yolk protein nitrogen rather than the onset of a caloric deficit. Egg size is also an important factor for yolk utilization since maintenance costs are directly related to tissue weight. Thus, because fishes produced from larger eggs are themselves larger and have correspondingly greater maintenance costs, they use their yolk less efficiently.

The influence of intraspecific genetic differences remains largely unexplored, yet it is probably a significant factor in yolk utilization. In a study with coho salmon *(O. kisutch)*, *Childs* and *Law* (1972) compared the embryonic development of progeny of normal males and normal females with progeny of precocious males and normal females. They found that offspring of precocious males developed and grew more rapidly and utilized their yolk reserves more efficiently.

B. Abiotic Factors

Interpretation of environmental effects on yolk utilization is complex. Growth during the endogenous nutrition period has been found to be reduced by extremes in pH sub- and supraoptimal temperatures, adverse salinities low dissolved oxygen concentrations, exposure to light, and exposure to sublethal concentrations of toxic xenobiotics. Within the tolerated range of each environmental parameter, decreases in utilization efficiency probably reflect increased costs of homeostasis and maintenance. As the upper and lower limits of each tolerated range are approached, however, deactivation of somatic synthetic systems can be expected.

Some environmental factors, particularly some xenobiotics, suppress activity and reduce the associated energetic costs, effectively freeing more yolk nutrients for growth and increasing utilization efficiency. *Leduc* (1978), for example, found the efficiency of yolk utilization in Atlantic salmon *(Salmo salar)* eleutheroembryos exposed to hydrogen cyanide (HCN) to increase with increasing cyanide concentrations in the range of 0 to 0.1 mg 1^{-1}. He attributed this effect to a reduction in activity at higher cyanide concentrations. Conversely, since embryonic activity circulates the perivitelline fluid, aiding gas exchange and ensuring distribution of the hatching enzyme, environmental factors that reduce activity might in some instances reduce yolk utilization efficiency and decrease hatching success. It is possible, therefore, for a particular environmental factor to either reduce or enhance the rate and efficiency of yolk absorption, depending on the exposure regime.

Temperature is probably the most variable environmental parameter affecting yolk utilization efficiency. The available data indicate that utilization efficiency reaches a maximum within the range of thermal tolerance of a given species; efficiency decreases toward both upper or lower limits of the tolerated thermal range. The exact shape of the curve describing the effect of temperature on utilization efficiency varies among species, probably in relation to differences in reproductive strategy and rearing habitat. By investigating both the behavioural and physiology responses to temperature, these authors were able to map the relationship between yolk utilization and temperature selection in California grunion *(Leuresthes tenuis)*.

NONYOLK NUTRIENT SOURCES DURING EARLY DEVELOPMENT

A. Putter's Theory

Putter's theory, the direct utilization of dissolved organic constituents of water by the early life stages of fish, has been largely discounted but deserves some comment. Although fish eggs are generally regarded as *cleidoic*, some embryos possess a limited ability to assimilate dissolved organic matter from the water. Embryos of a number of species [rainbow trout *S. gairdneri*; Atlantic salmon *S. salar*; Atlantic herring *C. harengus harengus* have been shown to take up and metabolize external substrates such as 14C-labeled pyruvate, acetate, glyoxylate and glycine. Uptake of exogenous substrates increases during embryonic development, reaching a maximum rate just prior to hatch. It is unlikely that exogenous substrates make a significant contribution to the general needs of developing oviparous fish, however, because of the paucity of dissolved organic matter in natural waters and the relatively slow transfer rates. *Siebers* and *Rosenthal* (1977) calculated that uptake of dissolved amino acids from a 2-μm solution provided only 1.1% of the energy requirements of developing Atlantic herring *(C. harengus harengus)* embryos.

Since juvenile fishes are capable of absorbing dissolved glucose and albumin via the gills and lateral lime system, assimilation of dissolved organic matter from the water can be expected after hatch. This is comparable to the nutrient absorptive role served by the external gill filaments of viviparous shark embryos. Assimilation of dissolved organic matter may in fact be important for the survival of species such as Pacific sardine *(Sardinops caerulea)* that encounter a metabolic deficit prior to acquiring the capability of exogenous feeding. *Wiggins et al.* (1985) proposed assimilation of dissolved organic matter as a possible reason for the low incidence of food ingestion in first-feeding larvae of American shad *(Alosa sapidissima)*. *Imada* (1984) found the growth of larval thalli *(Porphyra tenera)* was improved by addition of sugars and salts of some organic acids, especially arabinose, to the culture water.

B. Egg Membranes and Perivitelline Fluid

The nutritive role of egg membranes and perivitelline fluid is uncertain. Analyses of salmonid perivitelline fluid have demonstrated the presence of protein (25% of fluid wet weight), lipid (5-12%) , and carbohydrate (1-2%). Assimilation of these nutrients may be of some importance, especially prior to the formation of the yolk syncytium.

The external membranes of fish eggs contain carbohydrate, protein and lipid. *Smith* (1958) proposed that embryos assimilate nutrients released from the egg membranes by the action of the hatching enzyme during the hatching process. *Cetta* and *Capuzzo* (1982) found energetic evidence suggesting winter flounder *(Pseudopleuronectes americanus)* embryos utilize nutrients of the egg membranes. While a nutritive function of perivitelline fluid and egg membranes may be of some significance during certain segments of development, these materials do not represent substantial nutrient reserves when compared with yolk. This is attested to by the normal development of embryos in the absence of perivitelline fluid and egg membranes following mechanical or enzymatic dechorionation.

C. Viviparity

As defined by *Wourms* viviparity as *"a process in which eggs are fertilized internally and are retained within the maternal reproductive system for a significant period of time, during which they develop to an advanced stage and then are released."* This definition does not distinguish between ovoviviparity and viviparity. As such, viviparity in fishes can be seen to present an almost continuous progression, from a primitive pattern in which the egg contains sufficient yolk for complete embryonic development and the female provides only protection, to an advanced pattern in which the egg has little yolk and the embryo develops connections to maternal tissues at an early stage in order to satisfy its nutritional, respiratory, and excretory requirements. Nonetheless, all fishes with the possible exception of surfperches (embiotocids), whose eggs may lack yolk reserves rely on yolk nutrients for energy and growth during at least the initial portion of their early development.

D. Mixed Feeding

Most fishes studied under laboratory conditions are capable

of mixed feeding (exogenous plus endogenous) before incurring a metabolic deficit during the terminal phase of yolk absorption.

The available data demonstrate that mixed feeding offsets any potential metabolic deficit prior to complete yolk absorption and enhances growth and survival, especially during the terminal phase of yolk absorption. Early contact with food may also influence initial feeding behaviour resulting in increased food consumption and consequently greater larval growth. *Grigorosh* reported that, during mixed feeding, fish larvae with larger yolk reserves exhibited a diurnal feeding pattern while larvae with smaller yolk reserves fed continuously. He considered continuous feeding to be a disadvantage since it made larvae more prone to predation.

Mixed feeding has been difficult to corroborate in field surveys, perhaps due to a rapid rate of digestion, a low requirement for exogenous nutrients, diurnal feeding patterns, or defecation and regurgitation of ingested material upon capture. It is possible that to a certain extent the evidence of mixed feeding in laboratory studies represents an abnormal behavioural response to abnormal types and amounts of food under abnormal circumstances. Salmonids, in particular, exhibit a phase of "precocious" feeding when offered exogenous food from shortly after hatching. Ingestion of food during this phase does not benefit growth or survival when compared to unified controls. Precocious feeding may in actuality be disadvantageous, resulting in increased mortalities. *Ochiai et al.* (1977) observed that premature feeding by ayu *(Plecoglossus altivelis)* resulted in some fish swallowing food into the pneumatic duct of the swim bladder. Death ultimately ensued, apparently caused by bacterial and fungal infection of the swim bladder and adjacent viscera.

NUTRITION OF EMBRYOS AND LARVAE

The digestive and metabolic processes for first-feeding vertebrates are often undeveloped relative to those of juveniles or adults. It is known that the digestive physiology of eleutheroembryos is different from that of juvenile and adult fish. It is highly likely, therefore, that the nutritional requirements of early life stages are distinct from those of older fish. Moreover, since the liver and its complement of synthetase systems do not develop until some time after the yolk syncytium has been formed, prefeeding fish probably have a broader set of nutritional requirements than later life stages. Normally, this would not present a problem since the required nutrients would be provided by the yolk. Under certain circumstances (e.g., inadequate maternal diet), however, yolk reserves may be deficient in some essential component. The relationships among maternal diet, egg quality, and embryo survival warrant further research.

A major barrier to defining the nutrient requirements of prefeeding fishes has been the inability to alter the composition of the food resource, that is, the endogenous yolk reserve. In this regard, the use of defined media to rear embryos that have been separated surgically from their yolk reserves deserves further consideration. It may also be possible to remove the yolk, or a portion thereof, and replace it with a defined media. This would maintain the integrity of the yolk syncytium and minimize physical trauma to the embryo. Direct incorporation of radiolabeled substrates into the yolk sac using a replacement technique would eliminate the need for epidermal uptake, as used by *Terner* (1968), and would prevent maternal metabolism of the labelled substrates, as can occur when labels are incorporated into the yolk during oogenesis. Another potential method for defining the nutritional requirements of early life stages could be based on viviparous teleost embryos. By rearing viviparous embryos and eleutheroembryos on defined media it may be possible to determine their nutrient requirements.

6

Differentiation of Sex

In fishes there is a wide range of sexuality from synchronous hermaphroditism, protandrous and protogynous hermaphroditism, to gonochorism. The organism may be unsexual i.e., having separate male and female organs in separate organism of the same species. The term "bisexuality" used to denote hermaphroditism is likely to lead to a misunderstanding. This term should be used to describe gonochorist. The term *"intersex"* is used in the present review to denote either sporadically appearing or experimentally produced hermaphroditic individuals of a species in which all or nearly all individuals are *gonochoristic*.

Whether hermaphroditism is the more primitive condition from which *bisexuality* or *gonochorism* may have arisen or a specialization derived from the more usual vertebrate gonochorism is a matter for debate. The solution of this intriguing problem will require a great deal more information in the future. Nevertheless, fishes provide excellent material to approach the problems of sex differentiation and of evolution of sex among animals.

Unlike other vertebrates, a number of teleost fishes are hermaphrodites. *Atz* (1964), among others, defined the types of hermaphroditism. An individual is hermaphroditic if it bears both male and female tissues. If all, or nearly all, individuals possess both ovarian and testicular tissues, that species is hermaphroditic. *Synchronous* (balanced) *hermaphrodites* are those in which the male and female sex cells ripen at the same time, regardless whether or not self-fertilization is possible. In *consecutive* (metagonous) *hermaphrodites* there are two types: *protogynous hermaphrodites* that function first as females and then transform into males and *protandrous hermaphrodites* that transform from males into females. *Atz* lists 13 families of teleosts, belonging to five orders, that include species of these types. The transformation may be accomplished in several ways, depending upon the arrangement of sexual tissues.

TYPES OF HERMOPHRODITISM

Synchronous Hermaphroditism

Synchronous hermaphroditism is best seen in *Serranus scriba* (Serranidae) by Dufosse belong to this category. The gonad of Mediterranean serranids such as *S. cabrilla* and *Hepatus hepatus* is separated into the ovarian and testicular areas. *D'Ancona* suggested the possibility of self-fertilization. *Clark in S. subligerius*, proved that self-fertilization and development are possible. In captivity, an isolated individual can emit sperm and fertilize its own eggs. In both nature and in the aquaria containing two or more fish, the fish may form spawning pairs. Immediately after one fish in the female phase spawns, the partner fertilizes the eggs. At this moment, the colour pattern (vertical stripes) in the female phase changes to that of the male phase. Then, the first fish reverses its sexual role and acts as a male.

Of the four species studied by Smith (one belonging to the genus *Hypoplectrus* and three of genus *Prinodus)* are *synthronous hermaphrodites* while at least nine species are *protogynous hermaphrodites*. Most *Rivulus marmoratus*, an oviparous cyprinodont, are genuine hermaphrodites capable of internal self-fertilization. It is remarkable that over 10 uniparental generations have been propagated and each fish has been kept in lifelong isolation *ab ovo*. Tissue grafts between the parent and its offspring and among siblings were histocompatible, thus providing tight evidence of self-fertilization. According to *Harrington* (1967), low temperature (18°-20°C) tends to transform the hermaphrodites to males.

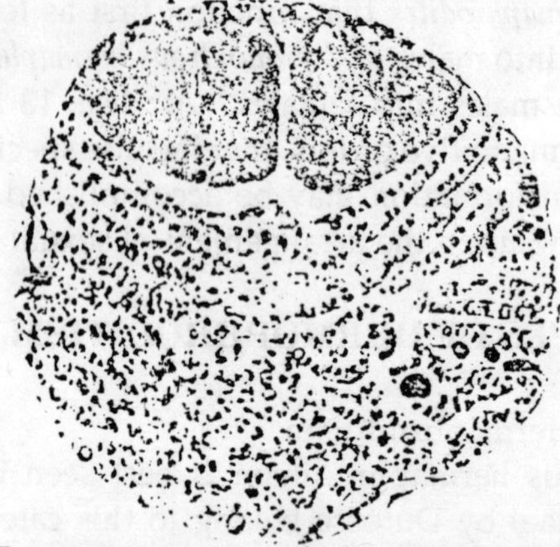

Fig. 6.1. Transverse section of the gonad of the synchronous hermaphrodite, *Serranus scriba* (Serranidae). Upper portion is the ovary and lower region the testis.

The members of families Serranidae, Sparidae, Centracanthidae, and Labridae are either *synchronous* or *consecutive hermaphroditism*. In all these families, however, there are some gonochoristic species.

CONSECUTIVE HERMAPHRODITISM

1. Protandrous Hermaphrodites

It has been observed that the Mediterranean bream. *Sparus auratus*, is a *protandrous hermaphrodite. Pasqua!* and *D'Ancona* described the precise process of gonadal differentiation from male and female phases. They demonstrated that its gonad consists of both testicular and ovarian areas from a very young stage. In smaller fish the lateroventral testicular region predominates over the ovarian zone, and in larger fish the reverse is true. In *Sargus (Dislodus) annularis* that there are some individuals in which the sexes are separated. However, *Pagellus centrodontus* is a *protandrous hermaphrodite.* Some sparids such as *Sparus aurctus, Sargus (Diplodus) sargus, Pagellus acarne,* and *Pagelus mormyrus* are also *protandrous hermaphrodites* while others such as *S. annularis, S. vulgaris. Puntazzo (Charax) puntazzo, Boops boops, Oblada melanula,* and *Dentex dentex are rudimentary hermaphrodites. Pagellus acarne* is protandrous.

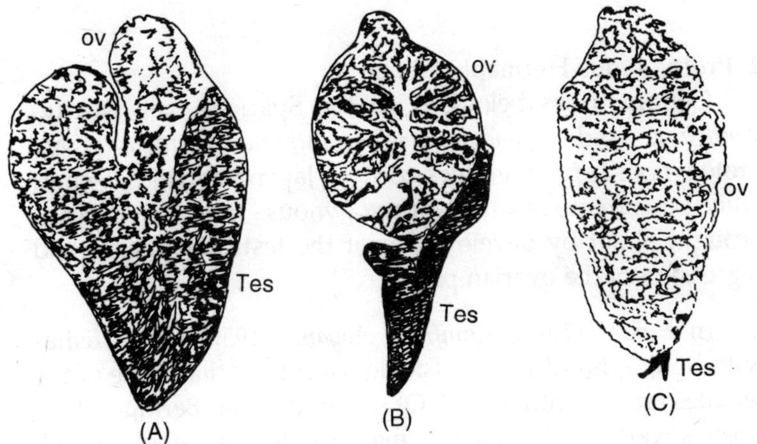

Fig. 6.2. Transverse sections of the gonad of the protandrous hermaphrodite, *Sparus auratus (Sparidae).* (A) Male phase, (B) transitory phase, and (C) female phase. Here Ov stands for ovary and Tes for testis.

In some species the *sex reversal* from male to female has been reported. The observed cases include *Acanthopagrus schelegeli* (syn. *Sparus longispinis, Mylio macrocephalus A. (Sparus, Mylio)latus*) and *Sparus sarba (Sparus aries, Rabdosargus sarba).* It was reported that not all individuals transform from male to female, i.e., some males retain maleness even when they become large. In the flat-head fish, Inegocia (*Cocialla*) *crocodila*, a large number of individuals (>1000) have been examined. Small individuals had testes and medium-sized fish hermaphroditic gonads with functional testes, whereas large individuals had ovaries. Another flat-head fish, *Inegocia (Suggrundus) meerdervoort,* is also a protandrous hermaphrodite.

In *Gonostoma gracile* a deep-sea luminescent fish, Kawaguchi and Marumo (1967) found that individuals less than 7 cm are mostly males and those more than 9 cm are invariably females. Sex succession takes place in the medium sized fish (7-9 cm), and hermaphrodites are frequently found in the specimens of 6-7 cm.

2. Protogynous Hermaphrodites

Certain fishes belonging to the Sparidae, i.e., *Pagellus erythrinus* and *Spondyliosoma (Cantharus) cantharus* are protogynous hermaphrodites. Among Japanese sparids, Dentex (*Taius*) *tumifrons* seems to be protogynous. Inversion of sex is brought about by development of the testicular region and regression of the ovarian part.

In *Dentex (Taius) tumifrons* Aoyama (1955) found fishes with hermaphroditic gonads. He concluded that some of the females change into males. Of serranids from Bermuda four species were synchronous hermaphrodites, as already stated, while at least nine species belonging to the genera *Epinephelus, Mycteroperca, Alphes, Petrometopon*, and *Cephalopholis* were protogynous hermaphrodites. Smith pointed out that there are three patterns of hermaphroditism in serranids : the *Serranus, Rypticus*, and *Epinephelus* type. The patterns are essentially similar to those hermaphrodites reported by Reinboth. *Epinephelus* and its allies have gonads in which the male tissue is present throughout the germinal epithelium lining the central lumen of the gonad. This male tissue becomes functional only after the female tissue has ceased to function. The genus *Rypticus* shows an intermediate type of gonad in which scanty male tissue is present in the lower part part but is also found intermixed with the female tissue. The protogynous *Chelidoperca hirundinacea* is also of this type. The eastern Pacific *Paralabrax clathratus* seems to be secondarily gonochoristic. Here, we can make an inference on the process in evolution of gonochorism from hermaphroditism. Figure 6.3 illustrates the three types.

Kuroda (1931) postulated that larger and red *Sacura margaritacea* and smaller and yellow *S. pulcher* are males and

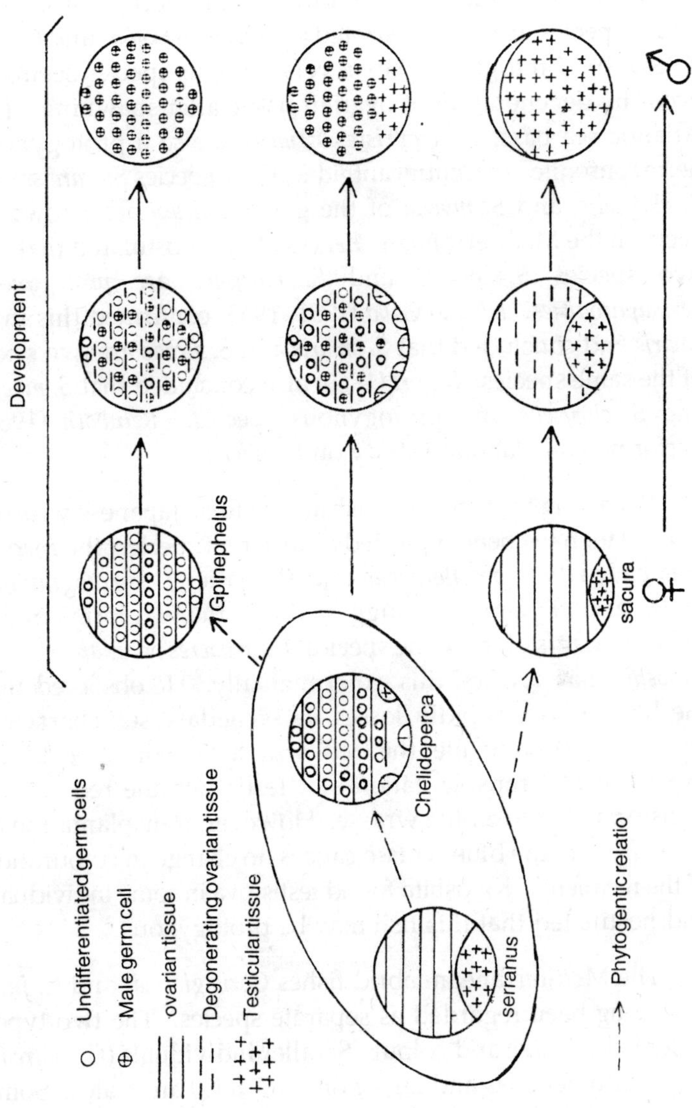

Fig. 6.3. Gonadal development and phylogenetic relations in serranid fishes.

females, respectively, of the same species. He found some intermediate individuals. The protogynous hermaphroditism in this species was demonstrated by *Reinboth* (1963) and *Okada* (1965a, b). Protogynous sex reversal in this species seems to occur by degeneration of ovarian tissue after spawning. The Atlantic sea bass, *Centropristes striatus* is also a protogynous hermaphrodite. In Centracanthidae, three species *Spicara smaris S. chryselis*, and *S. maena* of the genus *Spicara* are known to occur in the Mediterranean. *Zei* (1949) has postulated that the two species *S. smaris* and *S. chryselis* are *protogynous hermaphrodities*. *Lozano Cabo* (1951, 1953) confirmed this in *S. smaris* and concluded that *S. smaris* and *S. alcedo* are two sexes of the same species. *Lepori* (1960) also concluded that *S. maena* and *S. chryselis* and protogynous species. *Reinboth* (1962) performed an elaborate study on *S. maena*.

For a long time the greenish and reddish Japanese wrasses (Labridae) have been regarded as different species; the former was called *Julis poecilopterus* and the latter *J. pyrrhogramma*. *Jordan* and *Snyder* (1902) suggested that the two forms might be sex variants of the same species, *Halichoeres poecilopterus*. Y. *Kinoshita* has verified this experimentally. He observed that the blue wrasse rapidly losses its secondary sex characters after castration, while the red wrasse is not affected by ovariotomy. Transplantation of a testis into the red wrasse transforms it into a blue wrasse. However, transplantation of an ovary into the blue wrasse causes no change in colouration of the recipients. Kinoshita found testis-ova in some individuals and postulated that this fish may be protogynous.

The Mediterranean labrid fishes *Coris giofredi* and *C. julis* have long been regarded as separate species. The two types differ in both size and colour. Smaller individuals (*C. giofredi*) are usually females and larger ones (*C. julis*) are males. Some medium-sized fish show intermediate colour and have testes. The two types are the same species, *C. julis*, a protogynous hermaphrodite. An interesting fact is that a few large fish with *C. giofredi* colour are nevertheless males when their gonads are examined. Because each individual is at all times either a male or female, this labrid is termed a *"false gonochorist."*

The two species of labrids in the waters of the Mediterranean around *Livorno, Labrus turdus* and L. *merula*, are also protogynous species. In both species, all individuals less than 27 cm have ovaries. In the largest fishes of *L. turdus*, all individuals have testes. In *L. merula*, the largest individuals have a balanced sex ratio. This seems to indicate that while 50% of the fish change from female to male, 50% never change but remain females.

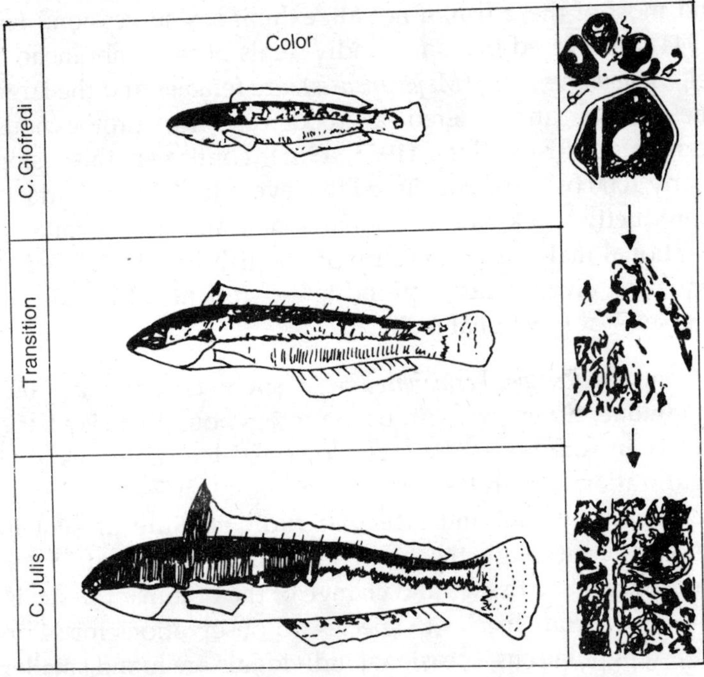

Fig. 6.4.   Change of colour and sex phase in the labrid, *Coris julis.*

Reinboth (1962) pointed out the existence of two types of morphologically distinct males in labrids (*Coris julis* and *Haliochoeres poecilopterus*). The primary male looks like a female but remains a male throughout its life, and the secondary male that changes to a male from the female. However, according to *Vandini* (1965), there are no primary males that retain the *C. giofredi* colour throughout their life cycle, viz., even the *C. giofredi* males eventually develop *C. julis* colour. In the stripped wrasse, *Labrus ossifagus*, there are some individuals with a red and others with a blue pattern. It was believed that the former is the female and the latter the male. *Lonneberg* and *Gustafson* (1937) reported that females greatly out number males in red fish and the reverse is true in blue-striped fish. They also found intersexual individuals changing from females to males. This seems to indicate that this form is protogynous and most of these fish, if not all, exhibit sex succession. *Liu* (1944) discovered that small individuals of the synbranchoid eel, *Monopterus albus (M. javanensis)*, are females and the large ones males and offered a good case for protogynous hermaphroditism. *Liem* (1963, 1965), confirmed this using nearly 1000 fish, and concluded that every individual starts its reproductive cycle as a functional female and then becomes a functional male. The sexuality of this fish has the following sequence; juvenile hermaphrodite —> functional female —> intersex —> functional male.

In *consecutive hermaphrodites* such as Sparidae and Serranidae, either protandrous or protogynous, there is a basic feature in common. The juvenile gonad has an ambisexual organization. From the very beginning of differentiation of the gonad, ovarian and testicular rudiments are present in every fish, although the topographical arrangement of the male and female tissues and change of the dominant tissue in time differ from species to species. In observations made on natural populations, occasional individuals are found smaller or larger than is usual for this sex. *Smith* (1959) noted unusually large females in some protogynous serranids, and *Reinboth* (1965) reported unexpectedly small males of *Centropristes striatus*, also a protogynous species. In *Dentex dentex* only some individuals undergo a transitional hermaphroditic stage. According to *Larraneta* (1964) about 5% of the fish in a population of *Pagellus erythrinus* are males throughout their life, while

45% transform from female to male and 50% never transform but remain females. *Rijavee* and *Zuvanovic* (1965) obtained comparable results in the same species. *Smith* (1967) proposed a theory of hermaphroditism in which he postulated that sex reversal takes place in different individuals at different sizes and ages viz., sexual succession is a prolonged continuous process for the population. "During the first time interval following sexual maturity, only a part of the population changes sex, and during each time period a fraction of the remaining individuals transform."

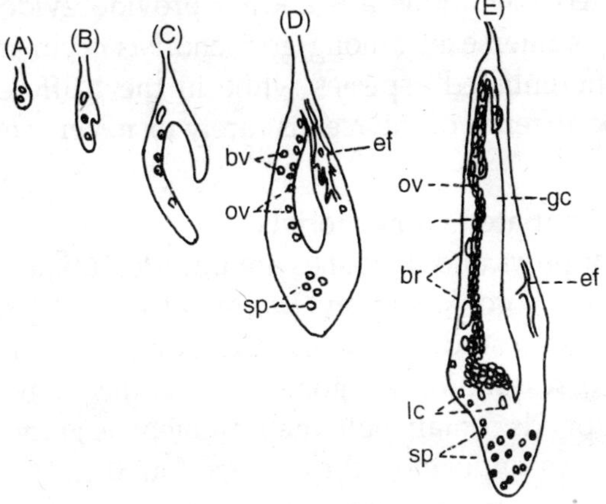

Fig. 6.5.   Schematic transverse sections of early developmental stages (A → E) of the gonad in Sparidae, where on stands for ovarigan spermatogonia; gc, gonocoelom; ef, efferent ducts; lc, lacunae; and bv, blood vessels.

GONOCHORISM

As stated already that *gonochorism* is a sort of bisexuality. Two types of species are known. In the undifferentiated gonochoristic species, the indifferent gonad first develops into an ovary-like gonad and then about one-half of individuals become males and the other half females. In differentiated gonochoristic species the indifferent gonad directly differentiates into either a testis or an ovary. In both types, sex differentiation seems to be brought about by male- and female-inducing substances. It is natural that undifferentiated species are more unstable than differentiated ones in sexuality. Unfortunately, however, the two types have been studied embryologically in only a few fishes. In the absence of embryological evidence, it is as yet hazardous to correlate the two modes with the occurrence of sporadic intersexes which represent a remnant of the embryological condition manifested late in some adults. However, a few species in which undifferentiated and differentiated conditions are known provide evidence that spontaneous intersexes among gonochorists occur mostly in the "undifferentiated" species while in the "differentiated" ones the occurrence of intersex is rarely or never seen.

A. Undifferentiated Gonochorists

The lampreys and hagfishes are undifferentiated species. Sexuality of the brook lamprey, *Lampetra lamottei (Entosphenus wilderi)* has been studied extensively. Ammocoete larvae up to 3.5 cm possess indifferent gonads. After this stage, some of the larval gonads contain both male and female germ cells. Of 15 adult males undeveloped ova were found in the testis of seven. In ammocoete larvae of *L. planeri* and *L. fluviatilis*, differentiation of the gonads is mostly completed at 5-6 cm in length. In *L. planeri* ammocoetes there is a small but significant excess of females. There are a number of transitional individuals with disintegrating oocytes and developing stroma, which may eventually become males. Among hagfish, Atlantic species, *Myxine glutrosa*, seems to be undifferentiated. It seems to indicate that this species is an intermediate type between hermaphroditism and gonochorism, in which juvenile hermaphrodites are common. The Japanese hagfish, *Eptatretus burgeri*, is also an undifferentiated gonochorist. Dean found only one male with an ovotestis out of 569 fishes.

Turning to the class Pisces, there are many undifferentiated gonochorists which normally develop into either male or females. However, sporadic intersexes are known to occur in these fishes. *Grassi* (1919) in the eel, *Anguilla anguilla*, and *Mrsic* (1923, 1930) in the rainbow trout, *Salmo gairdneri irideus*, have demonstrated that these species are undifferentiated. In the latter, accidental intersexes have been reported not infrequently. In Gadidae, sporadic intersexes have also been frequently reported. In the herring, *Clupea harengus*, there are several types of the intersexual gonad. In the minnow, *Phoxinus laevis*, *Bullough* (1940) found 10 fish with intersexual gonads. In all cases the ovarian portion was suppressed while the testicular region was normal. He thought that these intersexes represented transitory stages changing from female to male. The effects of an androgen and an estrogen supported this postulation. However, a complete sex reversal was not achieved with these hormones.

B. Differentiated Gonochorists

Only a few fishes have been shown to be differentiated gonochorists. Sexuality of differentiated species is fairly stable among fishes. *Wolf* (1931) demonstrated that the *"domesticated"* platyfish, *Xiphophorus* (formerly *Platypoecilus) maculatus*, is a differentiated species. *Bellamy* and *Queal* (1951) stated that not a single sporadic intersex has ever been found among 50,000 specimens. The situation is the same the medaka, *Oryzias latipes*. No spontaneous, true intersex has ever been found among more tan thirty thousand fishes studied during 40 years. *Yamamoto* (1953) showed that this species is a differentiated gonochorist. In this connection, Oka's report (1931b) on the occurrence of oviform cells in three males of this species may be mentioned. His finding was based on fishes which had been left to starve for 3 months. It is unfortunate and even strange that the title of this paper included the words *"accidental hermaphroditism"* since *Oka* carefully observed the cytological difference between these enlarged cells and genuine ovocytes, calling them *"pseudo-ovocytes"*. It is probable that these cells are enlarged proto- or spermatogonia prior to degeneration.

C. All-Female Species

There are certain species of viviparous toothcarp, *Poecilia (Mollienisia) formosa*, having hybrid origin. The *"all-female"* phenomenon in this form was first found by *C.L. Hubbs* and *Hubbs* (1932, 1946). In natural habitats, it propagates itself by mating with sympatric, related bisexual species, either *P. latipinna* or *P. sphenops*. However, paternal characters are not transmitted to offspring which are solely females. The monosexuality is the result of gynogenesis, the mate providing a stimulus to activate development of the ovum without syngamy. Nevertheless, there is strong evidence that the species is diploid. In terms of DNA levels, *P. formosa* nucleus is the same as in the diploid spaces *P. sphenops*.

C. Hubbs et al. (1959) discovered a single wild male in the Brownville population, and *Haskins et al.* (1960) reported a single male in their laboratory stock. It proved fertile in mating to *P. formosa* siring all females. The appearance of extremely rare males is still a matter for debate. In this connection, Kallman's study (1964b) deserves attention. He found among thousands of offspring from gynogenetic *P. formosa* mated with males of *P. vitatta* or *P. sphenops* 14 exceptional fish of which 12 exhibited paternal patterns and possessed morphologically intermediate features while the remaining two had mosaic patterns with some areas particlinous and others matriclinous. This shows that in rare cases syngamy occurs or a single chromosome from the male nucleus, governing pattern formation, may become accidentially incorporated into some or all cells of developing embryos giving rise to mosaic patterns.

Spurway (1953) claimed the occurrence of *"spontaneous parthenogenesis"* in two anomalous guppy females. Later this postulation was withdrawn and she assumed that the phenomenon results from the self-fertility of "functional" intersexes. Subsequent to Spurway's study, *Stolk* (1958) reported pathological gynogenesis observed in the guppy and the swordtail. The ovary of each female was infested by a phycomycete fungus, *Ichthyophorus hoferi*, which apparently was responsible for the pathological activation of the ova. More than 64 broods were produced by these fishes; these without exception were daughters. Stolk criticized Spurway's finding by saying that her fish might also be infected by the parasite.

The mode of reproduction of the two unisexual *"species"* or strains of the genus *Poeciliopsis*, viviparous toothcarps of northwestern Mexico, is unique to vertebrates. At first, it was considered that each of the two two undescribed species had both bisexual and unisexual strains. Bisexual strains were tentatively referred to as C and F and unisexual ones as Cx and Fx, respectively. *Miller* (1960) identified the C and Cx as *P. lucida*, but F and Fx are still indescribed. Later, *Schulz* (1966) showed that monosexual Cx is a form different from the bisexual *P. lucida* but closely allied to it. In nature, Cx propagates itself by mating with *P. lucids*. Hybridization experiments between Cx and males of two bisexual species *P. lucida* or *P. latidens* provided evidence that genetic factors of both parent combine to form the F_1 offspring. However, the entire male genome appears to be eliminated during ovogenesis, probably at meiosis. Hence, ova have no paternal genome in each generation.

Besides Cx, *Schultz* (1967) further reported two additional all-female stocks, Cy and Cz, which were previously thought to be Cx. *Poecilia lucida* provides sperm for these monosexual forms. While Cx and Cz are diploid expressing characteristics of both parents in the F_1, Cy is a triploid and in mating with males of various bisexual species, e.g., *P. latidens*, produces all-female, triploid offspring by gynogenesis devoid of paternal characters. Thus, the mode of reproduction of Cy is quite different from that of Cx and Cz in which paternal chromosomes of the F_1 generation are not transmitted through the ova, only those characteristics of the female germ line pass to the next generation.

Gynogenesis has been reported in certain natural populations of the crucian carp, called the *"silver goldfish," Carrassius auratus gibelio,* which produce all-female progeny. He regarded this phenomenon as a *natural parthenogenesis.* Artificial insemination revealed that a gynogenetic stimulus to the ova can be provided by the common goldfish, *C. auratus auratus,* or by the carp, *Cyprinus caprio.* In gynogenetic diploid forms, the diploid complement might be maintained by suppression of one of the meiotic divisions, reentry of the second polar body, or suppression of the first mitotic cleavage. The means by which the triploit Cy undergoes meiosis and produces fertile triploid eggs is obscure. At this point, it may be mentioned that *Rasch et al.* (1965) mated the gynogenetic, diploid *Poecilia formosa to P. vittata* and obtained offspring which were triploids as judged by their nuclear DNA. However, these triploids were sterile.

CONTROL OF SEX DIFFERENTIATION

A. Surgical Operation

Surgical castrations of female Siamese fighting fish, *Betta splendens,* have in rare instances produced fish in which the regenerated gonad was a functional testis. These experiments were first performed independently in the United States and later in Germany. The American authors obtained only seven reversals out of 150 spayings of which only three proved to be fertile, siring both sexes in mating with females. The German authors obtained three reversals, of which only one was fertile and fathered solely a total of 109 daughters. The latter experiment proved that the male is heterogametic.

B. Modification of Sex Differentiation by Sex Hormones

In the rainbow trout, *Salmo gairdneri irideus, Padao* reported that injection of a follicular hormone produced ova in the testes, while testosterone induced testicular tissue in the ovaries. *Ashby* reported the paradoxical similarity of action of estradiol and testosterone in the brown trout, *Salmo trutta;* both hormones retarded gonadal development and produced hypertrophy of the somatopleure. There was no evidence of sexual inversion. The result might be ascribed to the fact that he started hormone administration in alevins in which gonadal sex differentiation has already been established.

A number of attempts have been made is viviparous toothcarps to modify sex differentiation by administration of sex hormones, especially androgens, starting with newly born broods or adults. However, androgens mainly affect the secondary sexual characters and the action on the gonad has been either incomplete, pathological, or negative.

The effects of sex hormones have been studied. In the guppy *Poecilia reticulata.* The production of ovo-testis by administering *estrogens* to young male guppies was done by *Berkowitz.* He reported that the optimal dosage resulted in an almost complete reversal of testis to ovary. Miyamori produced ovo-testis by administration of *androgen* to young females. In the top minnow, *Gambuisa holbrookii,* which shows *"transitional intersexuality,"* *Lepori* produced testis-ova in females by administration of an *androgen* and in males by an *estrogen.* Comparable results were obtained in G. *affinis* by *Okada* (1944). In the swordtail, *X. helleri, Baldwin* and *Goldin* (1940) reported that *androgen* administration to young females not only simulated the male secondary characters such as formation of a caudal sword and the gonopod but also induced masculinization of the gonad after degeneration of ovocytes. *Querner* (1956) showed that *androgens* produce ovo-testis in genetic females. Contrary to the cases in ambphibians, corticoids have no effect on sex differentiation.

Vivian (1952a) reported that 11 incompletely hypophysectomized female swordtails developed an involuted ovary with so-called *"rest bodies"* and claimed that two or three of them showed partial masculinization. The fact that these rest bodies are not degenerated ovocytes but are cysts of a parasitic mycomycete has already been mentioned. This pathological change might be caused by infection of *Ichthyophorus hoferi.*

Failure to obtain complete reversal of sex differentiation in these studies may be because administration of hormones was started after the onset of gonadal sex differentiation. As stated before, although the guppy and swordtail are *"undifferentiated"* species, gonadal sex differentiation has already been established when broods are born Dzwillo's success (1962) in obtaining complete reversal in sex differentiation in the guppy.

In the oviparous toothcarp, *Oryzias latipes*, *Okada* claimed that the formation of testis-ova in adult males can be induced by administration of either *estrogens* or *androgens.* Treated males had gonads containing large oviform cells which look like ovocytes but which are not surrounded by follicle cells. *Okada* (1964b) obtained a true testis-ova by estrogen administration starting in juveniles. In this case, ovocytes are genuine since they are accompanied by follicle cells.

C. Complete (Functional) Reversal of Sex Differentiation

In the swordtail (undifferentiated species), *Dantschakoff* (1941) claimed to have induced males by the administration of *androgens* to females. If, however, this is true, it is possible that the induction was accomplished because this *"undifferentiated"* species is very labile in sexuality. As stated before this species has no sex chromosomes. By treatment with an androgen, female germ cells including ovocytes might degenerate and protogonia might differentiate into male germ cells.

In the differentiated species such as the platyfish and the medaka, where neither spontaneous sex reversal nor even sporadic intersexes occurs, induction of complete reversal may be impossible if hormone treatment starts after the onset of gonadal sex differentiation. The rank of stability in sexuality seems to be platyfish $\geq$ medaka > guppy > swordtail. No sex reversals have ever been successful by administration of sex hormones in the platies because heterologous sex hormones have inevitably been administered after birth when the gonadal sex differentiation has already been established.

The medaka, *Oryzias latipes*, (Fig. 6.9) an oviparous toothcarp, is a strict gonochorist. Generally, the sex-determining mechanism normal to this fish was established as XX for female and XY for male. Embryologically, it is a differentiated species (*Yamamoto*, 1953). Using our genetically analyzed strain (d-rR) of the medaka, in which white is female X^rX^r and orange-red is male X^rY^R, where R stands for xanthic pigmentation, functional reversal in sex differentiation has been accomplished in both directions by heterologous sex hormones. Thus, it has become possible to control sex differentiation *ad libitum* either from genetic males (XY) to functional females or from genetic females (XX) to males. In other words, we are able to inverse the prospective fate of the fish, genetically destined to develop into one sex and direct it to differentiate into the other. Success in reversing sex differentiation in both directions in the medaka rendered it possible to mate estrone-induced XY females with methyltestosterone-induced XX males.

In differentiated species, neither spontaneous intersex nor sporadic sex reversals can appear. Nevertheless, induction of complete reversal in sex differentiation can be achieved if the following necessary condition is fulfilled. *Sexogens (estrogens and androgens)* at suitable dosage levels should be administered consistently, beginning with the stage of the indifferent gonad and passing through the stage of sex differentiation.

NATURE OF NATURAL SEX INDUCES

A. Steroid Versus Nonsteroid Theories

The inducer theory of sex differentiation in amphibians by *Witschi*. Although the sex genes are effective sex determiners, their action appears to be mediated by the sex-gene-controlled sex inducers as far as the vertebrate is concerned. Terminology of sex inducers differs among authors.

There are *pros* and *cons* for the steroid theory of sex inducers. In amphibians, *Witschi* and his followers are *"cons"* and postulated a protein nature of the sex inducers without evidence. *Chieffi* (1965) also stated that *"the embryonic sex inducers,..... are in all probability different from sex hormones of the adults."* The postulation is based inter alia on (1) the so-called paradoxical feminization of the gonad by high doses of androgens, (2) the nonspecificity of sex hormones as sex inducers, viz., non-sex-hormonic corticoids are believed to also have the potency of sex inducers, and (3) the ineffectiveness of androgens on some WZ amphibians. These statement are based on the sex ratio scored by a limited number of gonads of young, the sex genotype of which is unknown. Neither a single case of androgen-induced functional nor a corticoid-induced sex reversal has ever been produced. Furthermore, criteria for ovocytes are questionable. Experiments on the adult male medaka show that not only estrogens but androgens, corticoids, and any noxious treatments such as heat treatment and starvation result in the formation of large-sized cells in the testis (testis-ova), without characteristics of ovocytes such as the presence of follicle cells. These oviform cells, which actually look like auxocytes, may be regarded as degenerating protogonia or spermatogonia enlarged prior to deterioration. As to point (3) above we should not rely on negative results. Reported experiments have been made by immersion in hormone. It may well be that androgen immersion cannot simulate the natural condition surrounding the protogonia and induce them to differentiate into male germ cells. Inducers are thought to be densely produced only by cells surrounding the protogonia.

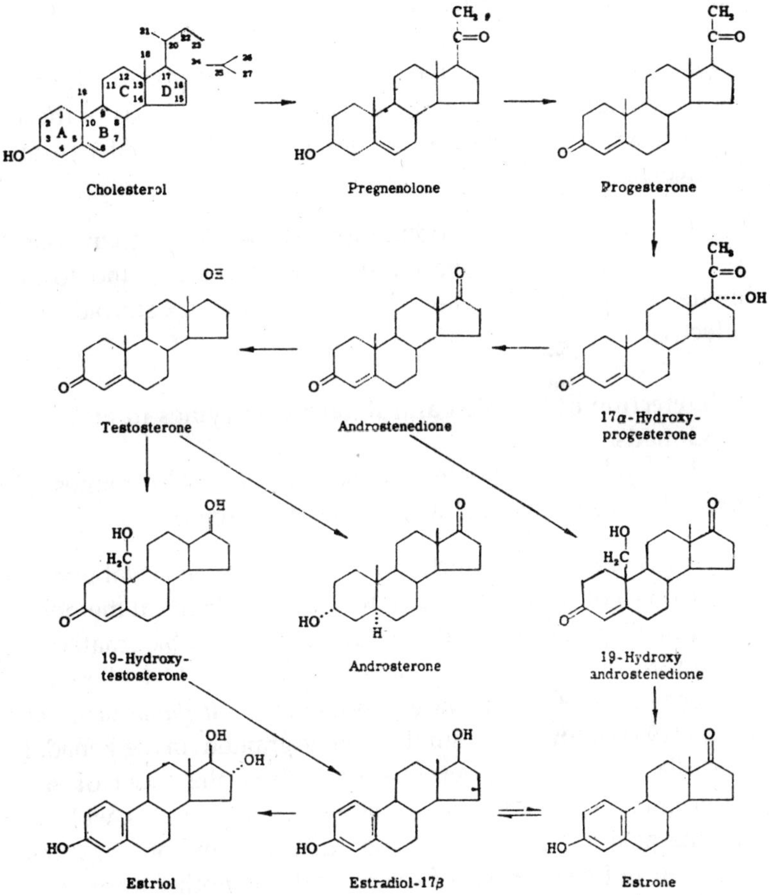

Fig. 6.6. A diagram of the current view of biosynthesis of key sex steroids.

In the *medaka* the following occurs : (1) *Estrogens* act as female inducers and androgens function as male inducers; no paradoxical phenomenon are engendered, (2) Sex hormones act specifically as sex inducers; non-sex-hormonic corticoids have no effect. (3) Effective dosage levels of estrone and methyltestosterone are so low that it may be possible that juvenile fish to elaborate a small amount of sexogens. (4) By administration of radioactive testosterone-16 ^{14}C propionate and diethylstilbestrol-(monoethyl-1-^{14}C), Hishide showed that these are selectively incorporated into the juvenile gonad. Point (4) seems to indicate that the sex-differentiating gonad requires more sex hormones than other organs. On the basis of these results, we are inclined to the sex steroid theory of natural inducers.

Of the naturally occurring steroids, pregnenolone, progesterone, and 17-hydroxyprogesterone are found to be ineffective while androstenedione and other sex steroids are effective

B. Detection of Steroids and Relevant Enzymes in Fish Gonads

It is yet to be known whether all the sex hormones of adult fishes are identical with those of mammals.

Research on the detection of steroids in the larval gonad at the onset of sex differentiation by classic chemical methods is difficult because of the paucity of available material. Therefore, almost all approaches depend on histochemical and enzymic method. In the elasmobranch *Scyliorhinus caniculas Chieffi* (1955) found a few sundanophilic granules in the gonadal medulla before sex differentiation. After the onset of sex differentiation, there appear many sudanophilic as well as Schultz-positive granules. The identification of enzymes involved in biosynthesis of sex steroids is another approach. Techniques have been developed for the histochemical demonstration of three relevant enzymes : ^{5}-3β-dehydroxysteroid dehydrogenase (Δ^5-3β-HSDH), ^{5}-3β-ketoisomerase Δ^5-3β-and 17β-hydroxysteroid dehydrogenase (17β-HSDH) which is DPN and TPN dependent. Together with ^{5}-3β-KIM ^{5}-3β-HSDH is involved in the pathway from pregnenolone to progesterone, one of the first steps of biosynthesis of sex steroids. *Chieffi* and co-workers (1961-1963) were unable to detect Δ^5-3β-HSDH in the embryonic gonads of the elasmobranchs, *Scyliorhinus caniculas, S. stellaris,* and *Torpedo marmorata*. The enzyme, however, was detected in the interrenal tissue of *S. stellaris* and *T. mamorada*. Bara (1966) showed the presence of 3β-HSDH in the testis of *Fundulus heteroclitus*.

Many questions remain unanswered. It is not known, for instance, where the supposed sex inducers are produced and in what way they influence the indifferent gonia (protogonia), to develop into either ovocytes or spermatocytes. Finally, the chemical nature of natural sex inducers still remains obscure. Research along these lines should be fruitful.

DIFFERENTIATION OF SECONDARY SEXUAL CHARACTERS

Our primary concern is with gonadal or primary sex differentiation. References to the manifestation of secondary sexual characters are so numerous that only a brief account is presented here.

It is convenient to divide secondary sex characters into two categories : (1) temporary characteristics which normally appear only during the breeding season such as nuptial colouration, pearl organs, and the ovipositor of bitterlings; (2) permanent organs developed fully at the onset of sexual maturity such as the gonopodium of viviparous cyprinodonts and the papillar processes on the male and fin and the urinogenital papilla in the female medaka, *Oryzias*. It is also convenient to define male- and female-positive secondary sex characters. The former is a character that is either specific to the male or is more developed in the male than in the female. The latter is the reverse. The mechanism by which these characters manifest themselves may only be elucidated by experiments.

Most of secondary sex characters of fishes are male-positive. *Kopec* was the first to demonstrate in the minnow *Phoxinus laevis* that the nupital colouration is dependent on the testicular hormone. This has been confirmed in the sticklebacks,

Gasterosteus pungitius and *G. aculeatus*. Similar results were obtained in the medaka *Tozawa* (1923) proved that the pearl organs of the goldfish, *Carassius auratus*, are controlled by testicular hormone. While castration results in the disappearance of these male-positive characters, ovariotomy causes no effect. This means that the absence of these in the female is not because of the inhibitory action of the ovary.

On the contrary, in the ganoid, *Amia calva*, *Zahl* and *Davis* (1932) showed that the gray-black pattern (male-positive) is absent in the female because of an inhibitory action of the ovary; ovariotomy produced this pattern. The male swordtail and the platyfish have complicated suspensorial skeletons including three hemal spines in the caudal vertebrae; these are necessary for proper functioning of the gonopodium. In the female, the three hemal spines are absent providing an undivided space required by the gravid female for the brood of embryos. *Vivian* (1952b) reported that spayed swordtails develop the three hemal spines as in the male, indicating that these are absent in the female by an inhibiting action of the ovary. *Rubin* and *Gordon* (1953) showed in the platyfish that α–estradiol benzoate induced the dissolution of one or two of these hemal spines in the male and methyltestosterone induced fusion of basal bony elements (*mesonost* and *baseost*) in the female, which are normally separated.

The secondary sex characters of the medaka, *Oryzias latipes*, have been described by *Oka* (1931a). *Nagata* demonstrated that castration results in a regression of both the shape and papillar processes of the male anal fin. Transplantation of a testis into the sprayed fish produces an almost perfect replica of the male characters. Administration of an androgen to females induces the male-type and fin.

In the topminnow, *Gambusia holbrookii*, *Dulzetto* (1933) noted the correlation between testicular development and the gonopod formation. A host of experiments on gonopod formation has been performed in the topminnow *G. affinis*, in the platyfish *X*.

maculatus, in the guppy and in the guppy-swordtail hybrid. Some of the earlier studies of hormone-induced characters, aimed at inducing modifications of sex differentiation by administration of steroids in poeciliids, have already been discussed X-ray irradiation in the females induced formation of a malelike anal fin has been reported by *Vivian*.

At this point, it may be remarked that not all fins are sensitive to *androgens*. For example, *androgens* have a stimulating action on the anal and dorsal fins of the medaka but exert no effect on the caudal fin. In the gobiid fish, *Pterogobius zonoleucus*. *Egami* (1959c) stated that while the dorsal fin is elongated by administration of an androgen, other fins are not affected. Another male-positive character is the large tooth in the male medaka. Large teeth can be produced in the female by administration of methyltestosterone. *Turner's* extensive researches (1960) demonstrated that regional parts in the anal fin of Gambusia have a differential sensitivity to androgens. The 3-4-5 ray complex is the most sensitive. He among others suggested that the enzyme pattern may be different in each area of the fin. Only a few female-positive secondary sex characters are present in teleosts. The urinogenital papilla (UGP) and the pelvic fin of the medaka and the ovipositor (elongated UGP) of the bitterlings are examples. Although the UGP is stimulated by both estrogen and androgen, it is far more sensitive to the former than the latter. It appears that normally the UGP of the female is produced by an *estrogen* from the ovary.

The UGP of the Indian catfish, *Heteropneustes fossils*, is also female-positive. *Kar* and *Ghosh* (1950) reported that injections of either an estrogen or an androgen in females resulted in drastic hypoplasia of the ovaries, oviduct, and the UGP. They attributed this seemingly curious effect of the estrogen to the paucity of gonadotropin through suppression of its action by a large amount of exogenous hormone and concluded that the UGP is under the control of the ovarian hormone.

The pelvic fin of the female medaka is longer than that of the male. *Niwa* (1965b) showed that this is because of an inhibitory action of the testicular hormone and cannot be attributed to a stimulating action of the ovarian hormone.

In the Japanese bitterling *Acheilognathus indermedia*, *Tozawa* (1929), showed that ovariotomy inhibits ovipositor lengthening. Research on the ovipositor formation in the European species, *Rhodeus amarus*, has provided a very confusing and puzzling problem. *Fleishmann* and *Kann* (1932) and *Ehrhardt* and *Kuhn* (1933) found that estrogen induces this reaction in the nonbreeding season. In 1935, American and German authors proclaimed that this response can be used for the diagnosis of pregnancy. It has been reported that urine from both men and pregnant women usually gives a positive reaction, while that from nonpregnant women may or may not cause this reaction. A male hormone in the urine was believed to be responsible. *Banes et al.* (1936) reported, however, that crude extracts of adrenal cortex gave a positive response while androsterone was negative. On the other hand, *Kleiner et al.* (1937) obtained a positive response by administration of androgens *De Wit* (1939) confirmed this and further stated that progesterone and deoxy-corticosterone (DOCA) and far more active than an androgen.

A survey of literature reveals that not only the above-mentioned substances but also *adrenaline* and *anesthetic* have been reported to be active in lengthening the ovipositer. *De Groot* and *de Wit* (1949) reported that alcohol, heat shock, and strong light also causes ovipositor growth. This would seem to indicate that stresses in the sense of *Selve* also cause the reaction. Furthermore, the presence of mussels and/or males and an exhalent current from mussels have the influence on the cyclic ovipositor growth.

7

Hatching in Fish

The process by which an animal changes its life from an *"intracapsular"* to a *"free-living"* type is called *hatching*. Among all animal groups, teleosts have been the most extensively studied. Hatching can be categorized into two types: *mechanical hatching* and *enzymatic hatching*. In the former, the egg envelope(s) is broken down, as can be seen in birds and in some insects, primarily by mechanical action such as a pressure exerted from within or mastication by the embryo. Similar types of egg envelope rupture have been reported in some aquatic invertebrates, although the evidence for participation of enzyme(s) is increasing. In the latter, the emergence of an embryo occurs after a preceding dissolution or softening of egg envelope by an embryo-secreted hatching enzyme. Hatching mechanisms of this type were first inferred and then observed in fish about 80 years ago.

In his studies on the development of the lungfish *Lepidosiren paradoxa Kerr* first described that the horny egg shell became quite soft so that the embryo could break it by a violent body movement. Although no experimental analyses were made at that time, he attributed this softening of the egg shell to a digestion by some ferment (enzyme) secreted by the embryo. Five years later, *Bles* (1905) also suggested that hatching of the amphibian. *Xenopus laevis* was due to an enzyme secreted from the frontal gland of the embryo.

Moriwaki found that at the time of hatching of *Oncorhynchus keta*, the inner layer of egg envelope was dissolved by the contents of perivitelline fluid. An undigested outer layer remained like a fragile veil that was then broken by the embryo. The contents of the perivitelline fluid derived from one embryo were so powerful that they could digest more than 15 egg envelopes at a temperature as low as 8°C. He concluded that the egg envelope — dissolving substance seemed to be a kind of ferment, although a strict identification was not accomplished. Furthermore, he found a large number of unicellular glands that become differentiated on the surface of embryonic body about 10 days before hatching, and he considered that the ferment must have been secreted from the mature glands only at the time of hatching, as the perivitelline fluid obtained before the time of hatching was inactive in dissolving the egg envelope.

Likewise, *Wintrebert* and *Bourdin* made extensive studies on the hatching of fish such as rainbow trout, goldfish, perch and other teleosts. They found that the movement of an embryo was not neccessary for hatching, as the embryo whose movement was inhibited with 0.03% chloretone was still capable of hatching. Perivitelline fluid obtained from the embryos just before hatching digested fertilized egg envelopes. Although they also noticed that a secretion from unicellular epidermal glands was responsible for the digestion of the envelope. *Wintrebert* (1912a) at first did not use the word ferment or enzyme for the digesting principle of rainbow trout embryos. He used this word for that of the perch embryos, *perca fluviatilis*. Thereafter, participation of *"ferment"* as the digesting principle became clearer, and the use of the term *"hatching enzyme"* was settled when *Needham* (1931) cited their work in his *Chemical Embryology*. Studies on hatching, though not many, were also made for various animal groups other than fish by the 1940s, and hatching enzymes had been described in aquatic vertebrates (fish and amphibian) and invertebrates such as ascidians, echinoderms cephalopods and insects.

HATCHING-GLAND CELLS

A. Differentiation and Maturation of Hatching-Gland Cells

In the early studies of fish hatching, it was observed that many unicellualr hatching glands appeared on the surface of embryos as they reached to the hatching stage. *Bourdin* (1926b) reported that hatching-gland cells were somewhat larger than other cells and contained many vacuoles, which were at first stainable with neutral red, but were gradually replaced by unstainable granules, while the concomitant mucous gland was an ordinary cell stained with mucicarmine. Bourdin regarded the hatching gland as being morphologically merocrine but functionally holocrine. This was confirmed later by many workers. Histochemical studies of hatching glands in *Oncorhynchus keta* were also reported by Inukai et al. (1939).

Differentiation of hatching-gland cells of the medaka *Oryzias latipes* was pursued histologically with light microscope and later with the electron microscope. According to *Ishica* (1944b), precursory hatching-gland cells in this species become visible around the pharynx of the embryos at the stage of eye pigmentation. One day before this stage (2-3 days after fertilization), the cytoplasm of some cells in the ventral endodermal cell mass becomes stainable with eosin and contain a small number of eosinophilic granules. At the stage of eye pigmentation, many giant cells containing eosinophilic granules are seen in the posteroventral region of the eye, and then only among the mass of endodermal cells. As development proceeds, the giant cells (-14 μm in diameter) migrate forward under the brain and begin to form the foregut.

Changes in histochemical stainability of hatching enzyme granules during development of fishes have been reported by several authors. According to *Ouji* and *Iga* (1961), developmental changes in the carp hatching gland can be classified into several stages. At first, the precursory hatching-gland cells contain a few granules stained faintly with acid fuchsin. The number of the granules increases gradually, and some of them become stainable with iron-hematoxylin rather than acid fuchsin. As development proceeds, the number of iron-hematoxylinophilic granules increases, until almost all granules are finally stained with this dye. In the case of azan or Mallory's stain, the granules are initially stained faintly with orange G. However, they become gradually stainable with aniline blue, though faintly at first, rather than orange G. In a well-developed gland cell, all secretory granules are stained deeply with aniline blue. Just before hatching, however, the granules become stainable again with orange G rather than with aniline blue. Thus, a secretory granule changes its affinity to dyes according to its stage of differentiation or maturation, having developmental stainability of a dual nature.

Localization of well-differentiated hatching-gland cells in fishes embryos differs from species to species. In the case of salmonid fishes, such as rainbow trout, the gland cells are distributed on the anterior surface of embryonic body and yolk sac, and on the inner surface of the pharynx and gill, while the distribution of gland cells in medaka is, as in some other cyprinodont fishes, generally confined to the inner surface of pharyngeal cavity. In most fish species, hatching-gland cells are distributed on the outer surface of embryonic body and or yolk sac, and are thought to be of ectodermal origin. In this connection, medaka is a rather exceptional fish, in the sense that the hatching glands are only in the pharyngeal wall and are of endodermal rogin. Also, in sturgeons, which have multicellular compact hatching glands, the gland cells are formed from an anterior part of gut and originate from the endoderm. The question of the germ layer from which hatching glands originate seems to be open to further study.

The hatching-gland cells of medaka can be distinguished from other endodermal cells early in development by their relatively large size, abundance of cisternae of endoplasmic reticulum, and a large electron-dense nucleus with a large nucleolus. At stages somewhat earlier than eye pigmentation, the secretory granules (hatching enzyme granules) appear first in the cytoplasmic matrix. The number of secretory granules in a gland cell increases markedly thereafter. Thus, hatching-enzyme synthesis in the gland cell seems to take place most actively around the stage of eye pigmentation, when the differentiating hatching-gland cells are increasing in size and forming a lining of pharyngeal cavity. In zebrafish embryos, *Brachydanio rerio*, the time of the first appearance of hatching enzyme granules coincides with that of eye pigmentation. A similar observation was also reported for rainbow trout. According to *Egami* and *Hama* (1975), hatchability of medaka embryos was remarkably decreased when they had been irradiated either with X rays (2 kR 250 R/min) or with γ rays (2 kR, 33.3 or 250 R/min) at the stages from optic vesicle formation to lens formation. Therefore, some irradiation-sensitive processes necessary for hatching-enzyme formation, such as mRNA synthesis, probably occur at these stages. In our preliminary studies on the isolation of hatching-enzyme granules, fraction 1 (600g x 10 min pellet) and fraction 2 (600-1000g x 10 min pellet) obtained from 0.3 *M* sucrose homogenates of embryos contained the secretory granules as they exhibited an ethylenediamine tetraacetic acid-sensitive (EDTA-sensitive) proteolytic enzyme. These secretory granule fractions obtained from day -3 as well as day-5 embryos exhibited a high specific activity of hatching enzyme, while those from day-2 and day-6 (posthatching) embryos showed almost no hatching enzyme activity. These results also indicate that the hatching enzyme is not yet formed in day-2 embryos, which correspond to the irradiation-sensitive stage, but it is synthesized soon after these stages. More recently, *Schoots et al.* (1982b) reported that the hatching enzyme could be detected immunohistochemically in hatching-gland cells of pike embryos, *Esox lucius*, at the 10- to 20-somite stage. Thus, it may be inferred that hatchinging enzyme synthesis in fish embryos is initiated in general just after lens formation but in advance of eye pigmentation.

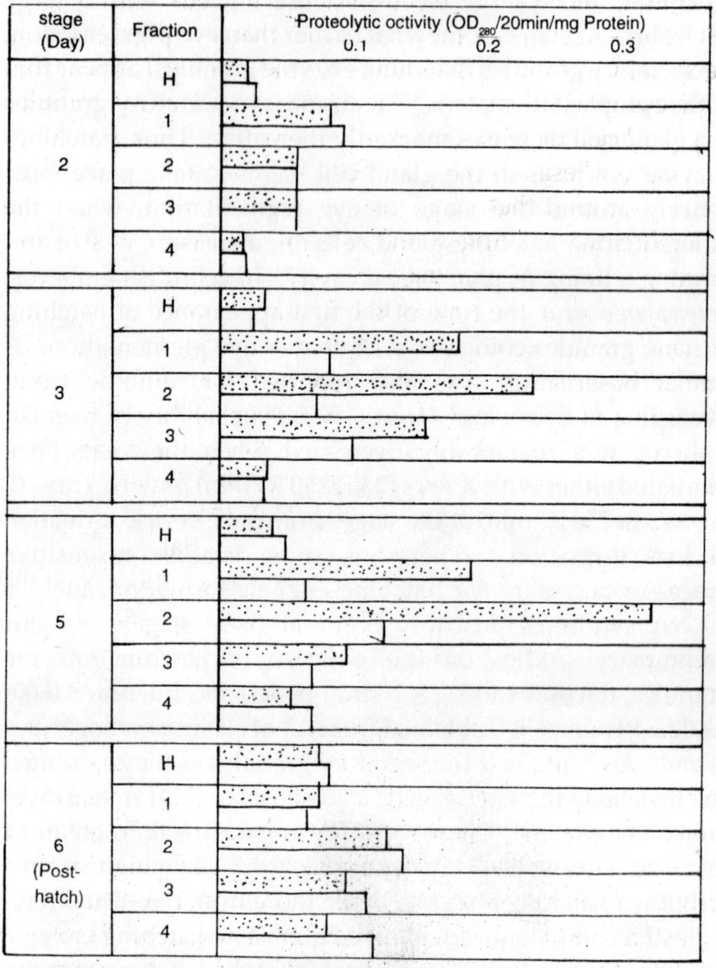

stage (Day)	Fraction	Proteolytic activity (OD_{280}/20min/mg Protein)
		0.1 0.2 0.3
2	H	
	1	
	2	
	3	
	4	
3	H	
	1	
	2	
	3	
	4	
5	H	
	1	
	2	
	3	
	4	
6 (Post-hatch)	H	
	1	
	2	
	3	
	4	

Fig. 7.1. Distribution of the hatching enzyme (EDTA-sensitive protease) activity among the subcellular fractions obtained from 0.3 M sucrose homogenate of medaka embryos at some developmental stages.

The hatching gland of medaka continues to produce secretory granules until nearly the prehatching stage. A few secretory granules found in the trans face of the Golgi apparatus were less electron-dense than most other granules, probably representing an immature state. Such immature granules could be found sometimes in day-5 embryos. In embryos close to the hatching stage, there were two types of secretory granules in hatching gland cells; one was homogeneouly electrondense and the other consisted of an electron-dense portion and a less dense portion. In the latter, the electron dense portion often took a crescent shape in the periphery of the granule, like a shell. Such heterogeneity of electron stainability in a hatching-enzyme granule has been seen also in some cyprinid embryos, *Brachydanio rerio* and *Danio malabaricus*, and salmonid embryos, *Salmo gairdneri, S. trutta, Salvelinus fontinalis*, and *S. pluvius*. As described above, histochemical stainability of a granule was reported to change markedly during development. Although it remains uncertain whether or not such a granule change is correlated with that of electron density, it is evident that the hatching-enzyme granules undergo some physicochemical changes during their maturation. A drastic change in the electron density of the granules in their last maturation phase seems to be closely related to the secretion process.

ULTRA STRUCTURE OF THE HATCHING-GLAND

1. Histological Studies

After being packaged in the secretory granules, the hatching enzyme is secreted into the periviteiline space, where it gains access to the egg envelope. In this section, the cellular and subcellular changes in the hatching gland associated with secretion will be discussed. There have so far been only a few studies on the cellular changes of the hatching gland during secretion. Several morphological changes have been reported in the hatching-gland cells in *Oryzias latipes* and *Odontobutis obscura*, respectively. In the former, the nucleus of the gland cell was invisible at the time of secretion and when secretory granules were released. In the latter, the nucleus remained in the gland cell, while the granules disappeared during secretion. *Yamamoto* (1963) reported that there were three types of secretory granules in the hatching-gland cells of medaka embryos. Type 1 granules were homogeneously electron-dense and were predominant at earlier development stages. Type 2 granules were as electron-dense as type 1 but contained a crescent-shaped shell of higher electron density. Type 3 granules contained somewhat granular contents with as low an electron density as the cytoplasmic matrix: they also had an electron-dense shell around the granular contents. The granules of this type were predominant in the embryos at later developmental stages. Just before secretion, a small hole appeared at the apical end of the cell, and type 3 granules seemed to be disintegrated within the cell.

2. Electrically Induced Secretion

It is sometimes difficult to predict accurately when the hatching-gland cells of an embryo initiate secretion under natural conditions. Several reagents or treatments have been reported that induce hatching-enzyme secretion in fishes, causing precocious hatching. Among them, an adequate dose of electric (AC) stimulation is quite effective in causing hatching-enzyme secretion in medaka as well as in rainbow trout. Rainbow trout embryos that would hatch normally about day 19-20 after fertilization at 15°C could be induced to hatch precociously on day 16-17 when they were stimulated with 100 V AC for 3 s 10 times with 5-min intermissions. In this case, hatching-gland cells on the surface of embryos became invisible a few minutes after the stimulation. When the dechorionated embryos were stimulated, hatching enzyme as determined by its caseinolytic activity increased in the medium.

Medaka embryos also hatch precociously upon electric stimulation. When cultured normally in a shaking incubator at 30°C, they hatch on day 6 if the day of fertilization was regarded as day 1. Natural hatching of control embryos begins early on day 6 and it takes almost one more day until all the control embryos complete hatching. However, stimulation of the embryos with 100 V AC for 5 s early on day 5 (~25 h earlier

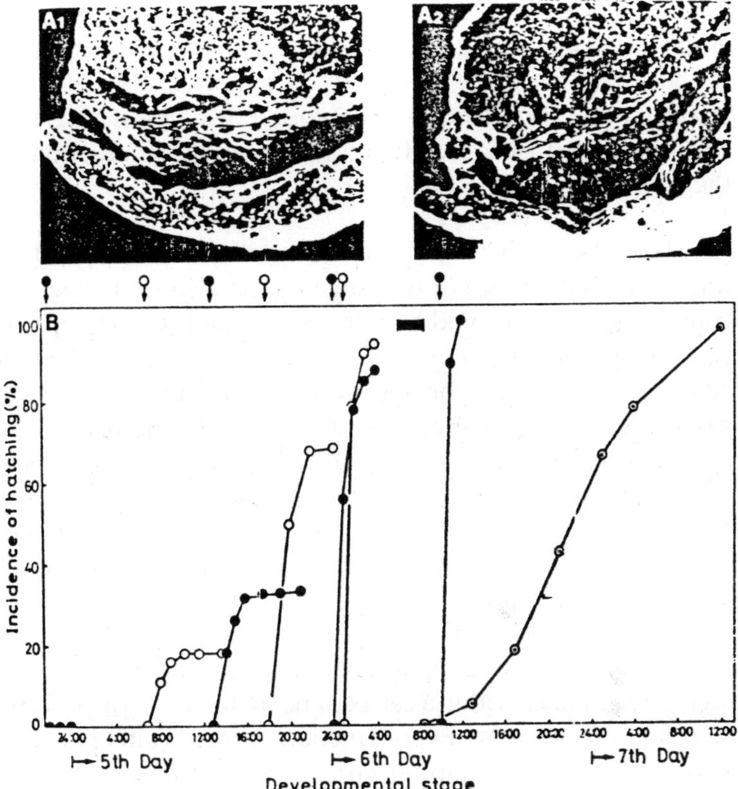

Fig. 7.2.   Induction of hatching enzyme secretion by electric stimulation in the medaka embryos. (A) Scanning electron micrographs (SEMs) of median cuts of the head of the embryos at the prehatching stage (A_1) before and (A_2) 5 min after stimulation. Many hatching-gland cells are seen as round protrusions in the buccal wall in (A_1) but not in (A_2) (x200). (B) Incidence of hatching of the stimulated (O, ●) or unstimulated (control, O) embryos. The arrow indicates the time of application of the electric stimulation (AC 100V, 5 s). A bold bar in the figure indicates the prehatching stage.

than the beginning of natural hatching) gave rise to precocious hatching of some embryos. Their responsiveness increased as development proceeded, and most embryos at prehatching stages could be induced to hatch. The gland cells of the embryos at the prehatching stage are considered to have been mature, in the sense that all cells were ready to secrete the hatching enzyme upon stimulation. It was found that almost all gland cells completed their secretion 5 min after the stimulation at the latest.

Exploiting this electric stimulation, a sequential ultrastructural change of the hatching-gland cells during the course of secretion could be followed in medaka. The gland cells are arranged side by side and are covered by a sheet of squamous epithelium on the inner wall of the pharyngeal cavity. Each epithelial cell has a hexagonal contour. Three adjoining epithelial cells meet at the apical center of each underlying gland cell. Just before electrical stimulation the gland cells were full of secretory granules of homogenous electron density, with the nucleus at the base. Near the Golgi apparatus, immature secretory granules with lower electron density were observed. Soon after the electric stimulation (usually ~30 s), a swelling of each gland cell occurred and the secretory granules within a cell became more clearly discernible as round protrusions. Every junction of three epithelial cells was separated and the apical surface of the underlying gland cell was exposed. Inside the gland cell, a coalescence of electron-dense secretory granules occurred to form a large mass of secretory substance surrounded by a limiting membrane. The contents of the coalesced mass appeared to be composed of fine granules, and its electron density was reduced remarkably. The electron density of the contents of a few uncoalesced granules, except for their peripheral part, was decreased slightly. As a result, these granules appeared to have a crescent-shaped shell of high electron density. The membrane surrounding the coalesced secretory mass became united with the cell membrane at the apex of the gland cell, forming an orifice through which the secretory substance flew out into the buccal cavity. It seems that the nucleus and cytoplasm including some endoplasmic reticuli still remained in some gland cells after secretion. About 24 h after the secretion, the openings at the epithelial junctions were reclosed and the open surface of the epithelium was flat, since any swollen gland cells were now absent underneath. However, some gland cells containing an electron-dense irregular-shaped nucleus and many fragmented cisternae of rough endoplasmic reticulum but no secretory granules were found to persist under the epithelium.

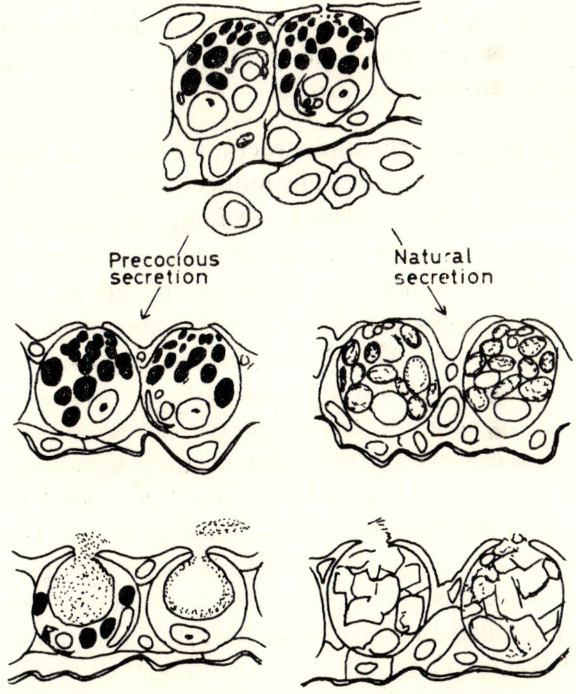

Fig.7.3.   Diagrammatic illustration of the ultrastructural changes in the hatching-gland cells of medaka embryos in the process of electrically induced precocious secretion and natural secretion

3. Natural Secretion

With artificially induced hatching, the hatching-gland cells in the process of natural secretion exhibited somewhat different features. Inside the gland cell, however, a different pattern of secretory-granule change was observed. The granules did not coalesce with each other and each granule became markedly electron lucent, except at its periphery. Thus, a hatching-gland cell just before natural secretion was, as observed in earlier studies, full of electron-lucent secretory granules bearing electron-dense shells. The granules became somewhat angular in shape, their membrane were dissolved partly, and their contents were mixed with cytoplasm before they were secreted from the cell. This process seemed to be different from that of exocytosis. The electron-lucent contents were composed of fine granules in this case also.

A comparison of the ultrastructural changes of hatching-gland cells during the electrically induced secretion with those during natural secretion shows two kinds of changes : those that are common to both types of secretion and those specific to each type of secretion. The common changes are swelling of gland cells, exposure of the apical center of gland cells following the separation of the epithelial junction, and reduction of electron density of secretory substance prior to secretion. By contrast, in natural secretion, no coalescence of secretory granules was observed, while in the induced secretion, many secretory granules of high electron density coalesced into a large mass of secretory substance and their electron density was decreased. A typical exocytosis was observed only in the induced precocious secretion, while the secretory granules were disintegrated and mixed with the cytoplasm of the gland cell in natural secretion.

In salmonid fishes, secretory granules become electron-lucent and fused together just before secretion. The gland cells discharge the granules together with some other cytoplasmic structures differently from ordinary exocytosis. After exhaustion of the secretory granules, the gland cells dissociate from the epithelium. However, according to *Schoots et al.* (1983a), there are three types of secretion in pike embryos: (1) exocytotic discharge via a secretion vacuole, (2) exocytosis at produced cell part, (3) intercellular exocytosis. Among them, type 1 is predominant.

Although the reason why such different types of secretion occur in the hatching gland is obscure, the fusion of secretory granules has also been reported in the process of secretagogue-induced secretion of various cells other than hatching-glands cells. For examine, in the rat peritoneal mast cell stimulated by the treatment with ferritin-conjugated sheep antibody to rat immunoglobulin (S and RIg-FT), an active degranulation occurs and the secretory granules coalesce into a large mass with low electron density. The membrane interaction in association with the degranulation leads to an exocytosis of the coalesced granular material. Thus, if seems that a fusion of granules occurs when the gland cells are forced to secrete somewhat rapidly by stimulants. Reduction of the electron density of secretory substances, irrespective of whether they are in granules or in vacuoles, may be related partly to hydration of the substances. According to recent work, the electron density of secretory granules remained high in the fully (or overly) matured hatching glands of medaka embryos whose hatching had been retarded by an "air-incubation". This fact strongly suggests that the electron-dense granules are already mature in the sense that they are ready to be secreted upon stimulation and that the reduction of electron density is not an indication of maturation but an indication of having entered into the secretion process. From the above results, it seems that some facets in the secretory changes, such as the increased fusibility of secretory granules, were manifested exaggeratedly in the electrically stimulated secretion of hatching gland compared to natural hatching. A stimulus for natural secretion may act somewhat more slowly or moderately, although its nature remains still uncertain. Even after natural secretion, some hatching-gland cells without any secretory granules but full of fragmented cisternae of endoplasmic reticulum persist under the epithelium. Similar persisting gland cells in the pike reportedly degenerated sooner or later by programmed death (apoptosis).

HATCHING ENZYME

Dissolution of the tough egg envelope by the secreted hatching enzyme is, together with the subsequent breakage of the remnant egg envelope (outer layer of chorion) by the embryo, a major feature of hatching in fishes. Thus, the nature of the hatching enzyme and enzymatic choriolysis have been foci of interest in the study of hatching. It is known that the hatching enzyme of fish has a proteolytic activity in addition to its egg envelope-dissolving activity *(choriolytic activity)*. Therefore, the hatching enzyme activity can be assayed tentatively for its *proteolytic activity*. Assay of the proteolytic (or peptidolytic) activity of the fish hatching enzyme has been performed using different substrates such as insulin casein or its derivatives, and some synthetic peptides. However, when a crude sample is used, the assay of only the proteolytic (or peptidolytic) activity is not appropriate for discriminating the real hatching enzyme from other concomitant proteases, if any. A turbidimetric method of semi-quantitative determination of choriolytic activity of medaka enzyme was devised to overcome such difficulty, although the method seems not be applicable to the rainbow trout enzyme. ^{14}C-labeled chorion was recently used as a substrate for *fundulus* enzyme.

The purification of hatching enzyme in fishes has been carried out since the early 1960s. In most cases, the enzyme has been obtained and purified from the hatching liquid, that is, the medium in which the embryos were allowed to hatch. *Kaighn* (1964) tried to purify the hatching enzyme the *chorionase of Fundulus heteroclitus* by gel filtration and sucrose density-gradient ultracentrifugation. The chorionase was sedimented between two molecular-weight markers, ribonuclease and hemoglobin, suggesting that its molecular weight was between 15,000 and 40,000.   Kaighn reported that the chorionase hydrolyzed tyrosine—threonine and threonine-proline peptide bonds in the B chain of insulin. Enzyme activity was inhibited with diisopro-pylphosphorofluoridate (DEP), a specific modifier of a serine residue.

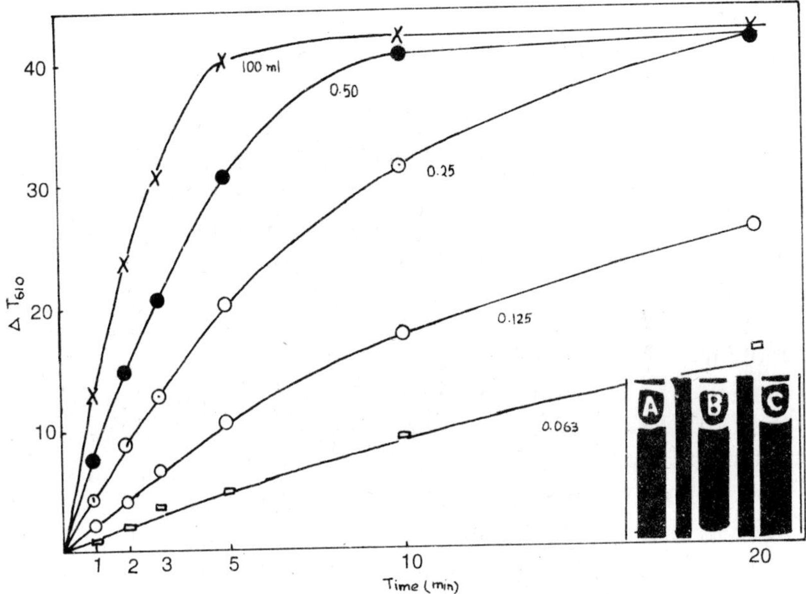

Fig. 7.4. Time course of choriotlytic activity of the medaka hatching enzyme as determined by turbidimetry. Inset shows the cuvettes containing the reaction mixture (A and B) with or (C) without chorion paste as substrate; A, after enzymatic digestion; B, before enzymatic digestion. ΔT_{610} refers to the increase in percentage transmission of the reaction mixture, including chorion paste, at 610 mm.

According to a recent report the swelling of the chorion seems to be an intermediate phase in the proteolytic digestion of the chorion. Assuming that this is true, choriolytic enzyme(s) obtained from hatching-gland cells of medaka separated as three different fractions moving toward the cathode at pH 8.6 on agar gel electrophoresis. Purification of medaka hatching enzyme used Sephadex column chromatography of the ammonium sulfate precipitates of hatching liquid, followed by CM-cellulose column chromatography, Sephadex G-75 column chromatography of the ammonium sulfate precipitate of hatching liquid gave two peaks of choriolytic and proteolytic activities, which were named PI enzyme (or enzyme I) and PII enzyme (or enzyme II), respectively. The specific activity of PII was much higher than that of PI. When the PII enzyme was fractionated by CM-cellulose column chromatography, a single peak of choriolytic and proteolytic activities coincident with a peak of protein was eluted by 0.02 M Tris HCi (pH 7.1)—0.3 M NaCl. Specific activities of this enzyme fraction (named PII-0.3) with respect to choriolytic activity and proteolytic activity were 212 and 183 times those of hatching liquid, respectively. The enzyme protein eluted as a single peak on Sephadex column chromatography and gave a single band moving toward the cathode on starch gel electrophoresis at pH 8.6 and 5.2 and on polyacrylamide gel disc electrophoresis (PAGE). However, this enzyme preparation showed some heterogeneity on sodium dodecyl sulfate (SDS) PAGE. Thus, it has yet to be determined whether the additional protein(s) in PII-0.3 are merely contaminants or some fragments of chorion protein associated with the enzyme. Recently, the secretory granules of the medaka hatching gland were isolated in 0.3 M sucrose. The aqueous extract of the isolated granules exhibited a high choriolytic activity, representing a single band of protein on SDS-PAGE. The molecular weight of PII-0.3 enzyme as determined by Sephadex column chromatography was reported at first to be about 8000. When determined on SDS-PAGE following *Weber* and *Osborn* (1969), however, it was about 21,000. The molecular weight of the enzyme in the aqueous extract of the isolated granules was also about 21,000 on SDS-PAGE. A similar discrepancy in the molecular weight was also reported for the pike hatching enzyme. This discrepancy may be attributable partly to a high affinity of the hatching enzyme for the supporting medium of gel filtration, and this method seems to be inadequate for the estimation of molecular weight of this enzyme. It seems highly probable that the hatching enzyme of medaka is a metal-loprotease but is not a serine protease nor sulfhydryl protease, as its activity is inhibited by ethylenediamine tetraacetic acid (EDTA) but neither by

DFP nor by iodoacetamide (IAM). Low concentrations of some monovalent and divalent cations activate the enzyme slightly, while high concentrations inhibit it.

It has been seen that hatching liquid of medaka embryos contains apparently two hatching enzyme fractions. PI enzyme (enzyme I) and PII enzyme (enzyme II). It was found, however, that a part of PI enzyme could be converted to PII enzyme (which was named enzyme PI-PII) through re-salting out and rechromatography on Sephadex. Such a conversion from PI to PII was observed if rechromatography of the PI enzyme was repeated. The properties of enzyme I, enzyme II, and enzyme PI-PII in terms of the sensitivity to some inhibitors were found to be almost identical. These observations strongly suggest that PI and PII enzymes were essentially the same enzyme, but that they behaved differently because of their different states of association with some heterologous substances such as hydrolyzed chorion. This view has been confirmed by our recent work. PI enzyme and PII enzyme have recently been highly purified from the hatching liquid by repeating Toyopearl gel filtration chromatography at pH 10. These procedures resulted in dissociation of the bound hatching enzymes. As a result, each PI enzyme and PII enzyme was found to consist of two types of proteases; one was a protease with high choriolytic activity (HCE) and the other was a protease with low choriolytic activity (LCE). The molecular weights are about 24,000 and 25,500 respectively, on SDS-PAGE.

Appearance of multiple hatching enzyme peaks on gel filtration chromatography has been reported also in some other fish species such as rainbow trout and pike. It seems that such a physical heterogenetiy is a characteristic of the hatching enzyme not only of fish but also of some other animal species such as sea urchin. An analysis of this problem will be useful for elucidation of the nature and the mechanism of action of this enzyme.

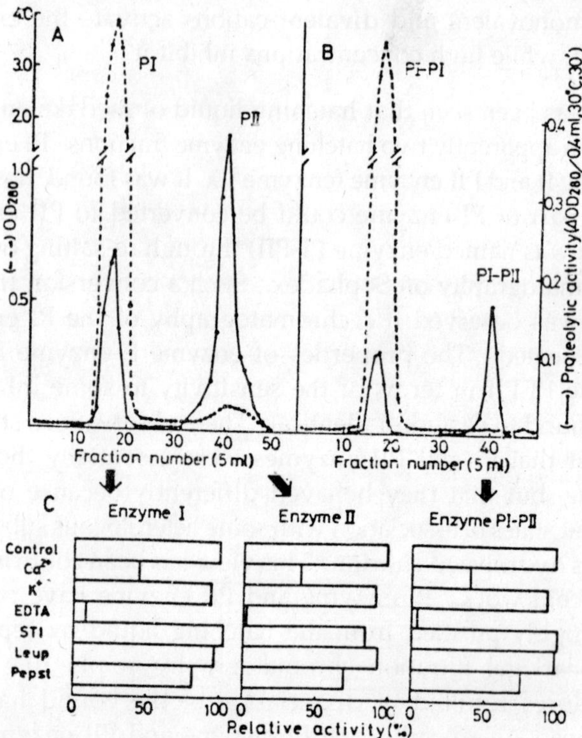

Fig. 7.5.   Fractionation of medaka hatching enzyme by Sephadex column chromatography. (A) Elution pattern of the ammonium sulfate precipitate of the hatching liquid on Sephadex G-75 column chromatography. (B) Elution pattern of rechromatography of the ammonium sulfate precipitate of PI through the same column as in (A). (C) Comparison of the properties of enzyme I, enzyme II, and enzyme PI-PII. STI, Soybean trypsin inhibitor (33 μg/ml); Leap leupeptin (33 μg/ml); Pepst, pepstation (0.33 μg/ml).

The hatching enzyme of rainbow trout was purified from the hatching liquid through the fractionation procedure similar to that of the medaka enzyme. The enzyme protein seems to be a basic protein from its behaviour on chromatography and electrophoresis. The molecular weight as determined by gel filtration chromatography was about 10,000. However, considering the probable inadequacy of the gel filtration method for determination of the molecular weight of hatching enzyme, it would be necessary to reexamine the molecular weight of the salmonid hatching enzyme with some other analytical methods. This enzyme also appears to be a metalloprotease, as it was inhibited by EDTA, ethyleneglycol bistetraacetate (EGTA), O-phenanthroline, or KCN, but not by phenylmethyl sulfonyfluoride (PMSF), tosyl-l-lysylchloromethane (TLCK), tosyl-l-phenylalanylchloromethane (TPCK), or iodoacetamide. It was reported that the activity of the EGTA-inactivated enzyme was restored only by from (fe^{2+}). The optimal pH of this enzyme was found to be around 8, resembling the medaka enzyme.

Recently, the hatching enzymes of *Fundulus heteroclitus* and of the pike *Esox lucius* have been well studied from biochemical and physiological viewpoints. Using chorion labeled with [^{14}C] iodoacetamide as substrate, *DiMichele et al.* (1981) examined some characteristics of *Fundulus* chorionase. This enzyme was found to be quite stable below 30°C, like the medaka enzyme and had a Q_{10} of 2.2 between 15 and 30°C. The pH optimum for the activity was between 8.0 and 8.5. This enzyme seems to be halophilic; in solutions of ions strength below 0.05 *M*, approximately 50% of the activity was lost in 18 h, but addition of NaCl within 48 h restored **the activity**. The optimum ionic strength was between 0.1 and **0.2 *M***. Such a salt requirement is seen in the enzymes of **medaka** and the marine fish *Gobius jozo. The Fundulus* enzyme was **found** to be insensitive to IAM but sensitive to EDTA as well as to PMSF. These results show that the *Fundulus* enzyme is a serine protease and/or metalloprotease but not sulfhydrylprotease. *Kaighn* (1964) also reported that *Fundulus* chorionase was **inhib**ited by DFP and was, therefore, presumably a **serine pro**tease. According to *Denuce* and *Thijssen* (1975), the hatching enzyme of zebrafish, *Brachydanio rerio*, also seems to be a serine protease.

Schoots and *Denuce* (1981) purified the pike hatching enzyme 1600 times from the original hatching liquid using affinity chromatography with carbobenzoxy-D-phenylalanyi-triethylenetetramine (Z-D-Phe-T) Sepharose. This enzyme is a glycoprotein containing 2% carbohydrate. The molecular weight of this enzyme was 10,000-15,000 by gel filtration but 23,500-25,400 with other methods such as PAGE, SDS-PAGE, and sedimentation analysis. The activity was inhibited by some metal chelators such as EDTA, EGTA, and *O*-phenanthroline but not by DFP, PMSF, iodoacetic acid, or *N*-ethylmaleimide (NEM). Furthermore, they concluded that this enzyme is a zinc metalloprotease based on atomic absorption spectrometry and renaturation experiments of the denatured apoenzyme.

B. Solubilization of Egg Envelope

Following activation or fertilization, the weak and fragile egg envelope of the unfertilized fish egg is transformed into a tough structure through a process called (water) hardening. The egg envelope (chorion) of the fertilized egg consists of a thin outer layer and a thick inner layer, the former being divided structurally into two sub-layers. The salmon egg chorion is composed of a scleroprotein, which was classified as pseudokeratin and was later named ichthulokeratin. There is a great similarity in amino acid composition of the chorion proteins among the eggs of samonids, *Fundulus*, and medaka, characterized by an abundance of proline and glutamic acid (and/or glutamine). The hardening occurs mainly in the thick inner layer. This tough structure protects the embryo against mechanical, chemical, and biological harm during development but also seems to be a barrier to the embryo in terms of hatching. Usually, the hatching enzyme is secreted shortly before an actual hatching occurs. In medaka, secretion occurs less than 1 h before hatching. Thus, the thick chorion is digested by the enzyme within 1 h or so, depending on temperature. We tried to stimulate the process of natural choriolysis in medaka eggs by incubating chorion pieces in the concentrated hatching liquid or the purified hatching enzyme solution (PII-0.2 enzyme) and to follow the sequential ultrastructural changes of the chorion pieces.

There are a number of honeycomb-like patterns on the outer surface of intact chorion. A large number of villi are present all over the surface of chorion, and many attaching filaments, much longer than villi, are restricted to the surface of vegetable pole area of the egg chorion. The inner surface of the intact chorion appears to be smooth, showing a somewhat parallel wavy pattern. On incubation of chorion pieces in the hatching liquid or in the purified hatching enzyme solution buffered with Tris-HCl at pH 7.2, the outer surface of chorion became rougher and many irregular dents and grooves appeared as the enzymatic erosion proceeded. In contrast to the outer surface, the inner surface of the chorion showed no irregular erosions during digestion; the partially digested inner surface remained smooth and flat. After complete digestion, a thin outer layer remained, apparently only slightly digested. When examined successively by transmission electron microscopy, the thickness of the inner layer was found to decrease evenly. It has been reported, however, that in the enzymatic choriolysis in vivo, instead of *in vitro*, the degree of inner layer digestion varied from fish species to species depending on the thickness of the inner layer of chorion. There are indications of enzymatic solubilization of inner layer at the peripheral (outer) parts, just beneath the outer layer. These areas seem to correspond to the dents of the grooves shown and to be caused by the enzyme that has permeated through the outer layer of chorion pieces incubated in the enzyme solution. The partially digested inner layer was slightly swollen, decreased in its electron density, and loosened into a fibrous network. The solubilized products of the inner layer could be fixed with glutaraldehyde and osmic acid, which suggests that the solubilized products were of high molecular weight.

In a preliminary experiment, it was found unexpectedly that the hatching enzyme digests of the medaka chorion contained a small amount of free amino acids as detected by thin layer chromatography. When a large number of chorions isolated from blastulae was incubated with the purified hatching enzyme (PII-0.3), most of them were digested to a clear viscous solution, leaving the outer layers with villi and attaching filaments undigested. The solubilized material was fractionated using Sephadex G-75 column chromatography into a major fraction (PI) of high-molecular-weight glycoproteins and a minor one (PII) of lower-molecular-weight substances, that is, small peptides and/or free amino acids. The former was fractionated further into two peaks of glycoproteins on Sephadex G-200 column chromatography: one named Fr. 1 was eluted at the void volume and the other (Fr. 2) eluted later. Both peaks are considered to be major constituents of the inner layer of the chorion. They were approximately equal in amount and were very similar to each other in amino acid composition as well as in absorption spectrum. Upon ultracentrifugal analyses, each of them exhibited symmetrical Schlieren profiles with sedimentation constants of 7.0 S for Fr. 1 and 4.5 S for Fr. 2. However, disc electrophoretic analyses revealed that Fr. 1 was highly heterogeneous, being composed of about six protein bands (C1-C6), while Fr. 2 was homogeneous.

Thus the major products of enzymatic choriolysis comprise about seven high-molecular-weight proteins. *Denuce* (1975) also found seven proteins including those of approximate molecular weight of 80,000 and 200,000 in the enzymatic hydrolysate of medaka chorion. On further examination of the pattern of Fr. 1 it was noticed that there seemed to be a regularity of chemical characteristics among the components of Fr. 1. After determining the molecular weights of the native forms of the Fr. 1 components, and of Fr. 2 following the method of *Hedrick* and *Smith* (1968), it was concluded that the net electric charge of each of the six components of Fr. 1 was approximately the same, but different from that of Fr. 2. The

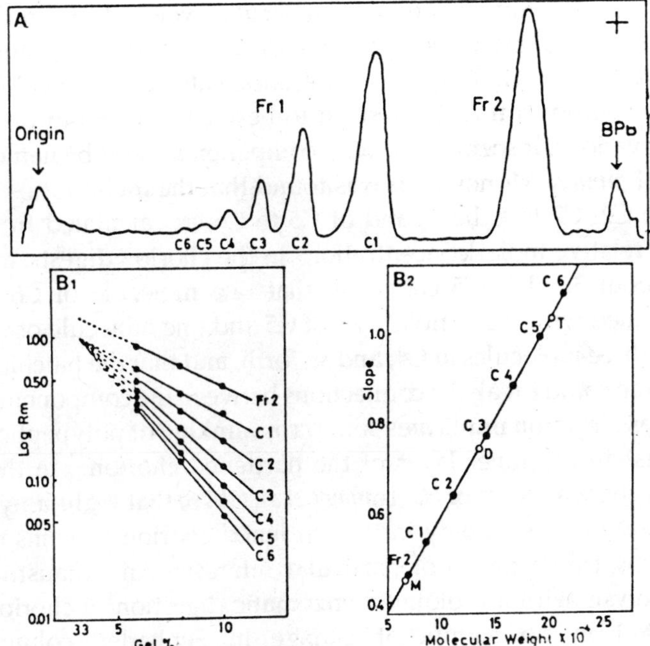

Fig. 7.6. Major glycoproteins solubilized from the medaka egg envelope by the action of the hatching enzyme. (A) Densitometric illustration of a polyacrylamide gel electrophoretic pattern. Major glycoproteins consist of Fr. 1 (C1-C6) and Fr. 2 (B) Molecular-weight determination of the six components of Fr. 1 and Fr. 2 according to the method of Hedrick and Smith (1968). The log R_m values of the glycoproteins were plotted against acrylamide concentrations (B₁), and their approximate molecular weights were estimated from the slope of the plots (B₂). M, D and T refer to monomer (MW = 67,000), dinner, and trimer of bovine serum albumin, used as references.

molecular weights of the six components of Fr. 1 ranged from 8.6×10^4 for C1 to 21.4×10^4 for C6 with an average molecular weight difference of about 2.6×10^4 between neighbouring components, while the molecular weight of Fr. 2 was approximately 7×10^4. It may be presumed that to the smallest component of Fr. 1, C1, is added a kind of repeating unit polypeptide of about 2.6 x 10⁴ molecular weight to form the second smallest polypeptide, C2, and to C2 is added the repeating unit to form C3, and so on. As described above, the pI values of all components of Fr. 1 seem to be identical. From these observations, it seems that these components could be named a Fr. 1 *family*. Moreover, it was found that the molar ratios of C1 to C2, C2 to C3, ..., and of C5 to C6, as calculated from their relative molar concentrations in the chorion digests, are all about 3. This might mean that one molecule of C6 is combined with three molecules of C5 and one molecule of C5 with three molecules of C4, and so forth, and that the hatching enzyme could break the connections between the components. This assumption implicates some cross-linking of polypeptide chains in the inner layer of the hardened chorion. In this connection, a report of *Hagenmaier et al.* (1976) that y-glutamyl-lysine was present only in the hardened chorion proteins of rainbow trout eggs is of particular interest. An exhaustive choriolysis with a prolonged enzymatic digestion of chorion resulted in no significant change in Sephadex column chromatographic pattern and PAGE pattern of Fr. 1 and Fr. 2.

These results suggest that the hatching enzyme digests the inner layer of chorion by hydrolyzing some restricted peptide bonds of its constituent proteins to give rise to two groups of soluble glycoprotein compounds, Fr. 1 and Fr. 2. Once these glycoproteins (C1-C6 of Fr. 1, and Fr. 2) are formed, they seem to be resistant to further enzymatic breakdown. A similar mode of choriolysis in principle may occur in the hatching of other fish species, although an accumulation of free amino acids is reported in the hatching fluid of rainbow trout. Some 40 years ago, *Hayes* (1942, 1949) suspected that the action of hatching enzyme was not hydrolytic, as the amount of amino-N produced by the enzymatic digestion of a capsule was so small. A limited cleavage of the inner-layer proteins of chorion by the hatching enzyme would give rise to the result compatible with the Hayes's observations as well as explain the efficient and rapid solubilization of chorion by the hatching enzyme.

C. Comparative Studies of Hatching

Although the hatching enzyme or enzymatic hatching was documented first in fish, hatching has been described in many other animal species, and the number of such examples is increasing. Whether there is a phylogenetic correlation to the mechanisms of the enzymatic hatching in various animals is still uncertain, but it would be useful for better understanding of fish hatching to make reference to the enzymatic hatching in other animal groups. In this section a brief survey will be made of hatching in amphibians and sea urchins. Discussion will be extended to the digestion of the cocoon by cocoonase in insects and the solubilization of the vitelline envelope by sperm lysins, as these phenomena are closely related to enzymatic hatching in some respects.

HATCHING IN AMPHIBIA

Hatching in Amphibia has been studied in many favours. In the *Xenopus laevis* the role of frontal glands in hatching has been studied. The hatching gland of Amphibia is of ectodermal origin. Although there was a view that the anuran hatching gland originated from the neural crest, it was found recently that most gland cells were derived from the superficial epidermal cells situated on the neural crest. Moreover, *Yoshizaki* (1979) succeeded in inducing the hatching-gland cells from the explanted superficial layers of the presumptive ectoderm in *Rana japonica* with LiCl. Thus, it is believed at present that the anuran hatching glands originate mostly from ectoderm other than the neural crest. In pilocarpine-induced secretion, the electron density of secretory granules of gland cells decreases and a partial coalescence of some granules occurs. As to the escaping of anuran embryos from the jelly layers, there have been some reports suggesting a nonenzymatic process. In the toad *Bufo vulgaris formosus*, there are four jelly layers, which are named A, B, C and D, respectively, from the outer to the inner layers. A and B from a jelly string and the innermost D interface with the vitelline envelope. When the embryos attain the late neurula stage, they escape preliminarily from the jelly string by perforating layers A and B, while each embryo remains still covered by layers C and D. Hatching from these layers occurs only when the embryos reach the tail-bud stage. Escape from layers A and B is not due to any proteolytic action but is primarily due to swelling of layer C. *Kobayashi* (1954b) argued that an augmented respiratory activity of the embryos was closely related to the swelling of layer C. Thus, enzymatic hatching is preceded by a nonenzymatic process. A similar observation was also made on *Xenopus laevis* embryos by *Carroll* and *Hedrick* (1974), who reported that the hatching process consisted of two temporally distinct phases, that is, phase 1 and phase 2. In phase 1, the embryo escapes from the outer jelly layers, J3 and J2, without the aid of a hatching enzyme, but probably by a physical process such as water inhibition by the inner jelly layer, J1; in phase 2, a hatching protease participates in the dissolution of the vitelline envelope. This two-step hatching process seems to be of some interest and suggests that such an analysis should be made also in fish hatching, although no thick multijelly layers are present. In salmonid embryos, the hardness of the egg envelope gradually decreases long before actual hatching. It seems improbable that the hatching enzyme had already been secreted and participated in such envelope softening. Thus, there is a possibility of participation of some factor(s) other than the hatching enzyme in a preliminary softening of the egg envelope in the hatching of some fish. The amphibian hatching enzyme is also a protease. The *Rana Chensinensis* enzyme was purified about 100-fold from its original culture medium. The molecular weight is approximately 55,000-60,000 and its optimum pH is 7.4-7.8. This enzyme is not affected by Na^+, $Km,^{2+}$ or soybean trypsin inhibitor but is strongly inhibited by Ca^{2+}, Mg^{2+}, EDTAA, and DFP. The *Xenopus laevis* hatching enzyme was purified 2200-fold over the starting crude hatching media. This enzyme has two enzymatically active charge isomers present with molecular weights of 62,500. The activity toward its natural substrate is optimal at pH 7.7. The enzyme is inhibited by Zn^{2+}

and by EDTA and seems to be a serine protease from inhibition by DFP and PMSF. From these characteristics, the amphibian enzyme is different from the enzymes of Oryzias and salmonids but somewhat similar to those of *Fundulus and Brachydanio.*

HATCHING IN ECHINODERMS

The mechanism of enzymatic hatching has been best studied is the sea urchin. The echinoderm hatching enzyme was first documented in *Strongylocentrotus (Hemicentrotus) pulcherrimus,* by *Ishida* (1936), and its properties were studied by *Sugawara* (1943). The optimal pH of its proteolytic activity was around 8.5-9.5, and low concentrations of Ca^{2-} seemed to be necessary for its activity. Of all the animal species, purification of hatching enzyme was tried first in echinoderm by *Yasumasu* (1961), who obtained *Anthocidaris crassispina* enzyme in crystalline form. The optimal pH of the crystalline enzyme activity was 8.2-8.4, and nearly a half of the activity remained even after heating at 60°C for 10 min. Following these pioneering studies, there have been many studies of purification and partial characterization of hatching enzymes from various species of echinoderms. The hatching-enzyme characteristics are general in some respects but contradictory in others. All the echinoderm hatching enzymes reported heretofore are found to require a suitable concentration of Ca^{2+} for their maximum activity, while the effect of Mg^{2+} on the activity is not settled. The enzyme seems to be retained on DEAE-cellulose at pH 8.0 or 8.2. There has been considerable variation in the reported molecular weights. On the one hand, the *S. purpuratus* enzyme was purified and the molecular weight was reported to be about 28,500 and 30,000 on SDS-PAGE and Sephadex column chromatography, respectively, on the other hand, the molecular weight of the *S. intermedius* enzyme, which was purified to a single band on SDS-urea-PAGE and separated from concomitant b-1-3, glucanase, esterases and most proteinases, was found to be about 44,000 on SDS-urea-PAGE and 45,000 on Sephadex column chromatography. This variation in molecular weights seems to be attributable in part to a physical heterogeneity of echinoderm hatching enzymes in the original hatching liquid; the enzyme may be combined with some heterologous molecules such as various-sized materials of fertilization envelope. Moreover, *Nakatsuka* (1985) reported recently that the sea urchin hatching enzyme was present inside the blastula in proenzyme form, which had a larger molecular weight.

Another line of hatching enzyme study in echinoderms is of the genetic control of this enzyme protein. *Koshihara* and *Yasumasu* (1966) reported that the *Hemicentrotus pulcherrimus* enzyme could be synthesized *in vitro* using chromatin of the embryos about 3 hr before hatching as a template. *Barrett* and *Angelo* (1969) reported, however, that the echinoid hatching enzyme was entirely maternal based on their studies on reciprocal hybrid embryos, whose parent echinoid species, S. *purpuratus* and S., *franciscanus*, had the hatching enzymes of different sensitivities to the added Mn^{2+}, *Showman* and *Whiteley* (1980) have recently reported that the messenger RNA of the echinoid hatching enzyme is newly transcribed in advance of hatching, based on their well-devised experiment using the hybrid andromerogons between the two echinoid species S. *purpuratus* and *Dendraster excentricus*. The developmental stage at hatching is much earlier in echinoids than in fish; the echinoid hatching enzyme seems to be one of a few specific proteins that may be synthesized during cleavage. Therefore, it seems highly probable that the fish hatching enzyme is also synthesized under the control of not the "*maternal*" genome but the embryonic genome. It has not yet been observed electron miscroscopically that the echinoid hatching enzyme is packaged in any particular structure such as secretory granules, although *Nakatsuka* (1985) reported that a granular hatching enzyme could be obtained by centrifugation.

Other Phenomena Related to Hatching

There are some enzymes similar to the hatching enzymes in a strict sense, that is, an embryonic enzyme dissolving a fertilization envelope. The best studied among them is cocoonase, which is synthesized in and secreted from the maxillary galea of the pupa of certain saturniid moths and participates in the digestion of the cocoon, making the "escape hatching" of the pupa possible. This enzyme is an organophosphate-sensitive protease, resembling trypsin in its substrate specificity, amino acid composition, and molecular weight (~24,000). Cocoonase is synthesized in zymogen-producing cells of the galea and transported into zymogen-strong vacuoles. A remarkable characteristic of this enzyme is that the active enzyme is deposited on the galea as a semicrystalline encrustation after secretion. The enzyme powder is dissolved in a galeal exudate, which serves as the buffer solvent for the enzyme before being applied onto the inner surface of cocoon. Thus, it seems that a natural enzyme solution can be easily obtained from a pupa just before *"escape hatching."* The second feature of this enzyme is its unique mechanism of action; this enzyme digests the cocoon by hydrolyzing not its main constituent protein, fibroin, but sericin, which glues the fibroin fibers together.

Another group of egg envelope-dissolving enzymes (or agents) of interest in comparison with fish hatching enzymes is the so-called egg membrane lysin or *sperm lysin*, although it is quite different from the hatching enzyme. Many lysins have so far been described in vertebrates and invertebrates. The lysin is thought to be localized in the *sperm acrosome* and to participate in the dissolution of the egg envelope when the sperm penetrates the envelope. Mammalian *acrosin* is one of the best-characterized vertebrate lysins and is similar to *trypsin*. It is assumed that mammalian *acrosin* is functional in a form bound to the acrosomal membrane under natural condition. There have been many studies on the egg envelope dissolution by sperm lysins in marine invertebrates. Among them are some reports in which the lysins of some gastropod sperm are not enzymes but rather low-molecular-weight proteins, which dissolve or loosen markedly the vitelline coat of eggs by combining with it *stoichiometrically* to form a soluble complex. It seems improbable that such a nonenzymatic action of sperm lysins is prevalent in marine invertebrates; it appears that lysins of this type are found only in some restricted animal groups such as archaeogastropods. However, the mechanism of action of this gastropod lysin gives us important information about a facet of the mechanisms of egg envelope dissolution or of the biological breakdown of a noncellular structure composed of scleroprotein. It seems that the hatching enzyme also has a high affinity for its natural substrate. Is it unreasonable to think that the mechanism of action of the archaeogastropod lysin is an extreme example of the interaction between the egg envelope-dissolving factor and its substrate ?

PHYSIOLOGY OF HATCHING IN FISHES

Factors Controlling Hatching in Fishes

As stated already hatching of fishes is a developmental stage-specific phenomenon. In fact, the embryo must have attained a particular developmental stage and have fully matured hatching-gland cells before hatching occurs. However, attainment of a specific developmental stage is not sufficient to cause actual hatching. Some triggering stimuli either extrinsic or intrinsic, have to be received by the appropriately developed embryo in order to induce hatching enzyme secretion. Thus as *Smith* (1957) pointed out, the onset of hatching in teleosts is a complex phenomena. There have been many factors or treatments that are reported to either stimulate or suppress the hatching of fishes. They are believed to influence the secretion of the fish hatching enzyme. In one of the earliest studies of the factors inducing fish hatching, *Armstrong* (1936) argued that there were two factors involved: the lashing movement of the embryonic tail and the secreted hatching enzyme. He showed that no hatching occurred when either of these factors was inhibited. At present, it is well-known that the lashing movement of embryo is effective only after the enzyme has exerted its digesting action on the inner layer of the chorion.

1. Oxygen Availability and Respiratory Movement

Ishida (1944b) observed that in medaka embryos, an opercular movement took place followed by disintegration of the hatching gland shortly before hatching. When the opercular movement was suppressed by treatment with 0.25 M KCl, the gland did not disintegrate. On the other hand, the breakdown of hatching glands occurred when the embryo was treated with 0.1 or 0.2% Vernoal-sodium, which affected the whole body movement but not the opercular movement. It was further observed that the gland cells could be disintegrated by water flow from a capillary that had been inserted into the pharynx of the embryo. Thus, the enhancement of opercular movement of embryos seems to be one of the phenomena most closely correlated with the initiation of the hatching enzyme secretion in medaka, although it remains obscure whether or not water flow is the sole cause for the hatching-enzyme secretion. When the shaking of a large number of medaka embryos was stopped just before hatching and they were heaped up in a small beaker, their hatching was markedly accelerated. Such treatment is considered to cause oxygen shortage, which in turn stimulates the respiratory activity of the embryos. Beside these observations, the results of a close relationship between respiratory activity of the embryo and hatching have been accumulated. According to *Trifonova* (1937), hatching of salmon embryos can be accelerated if they are subjected to asphyxia by bubbling hydrogen through the hatchery water. Similarly, hatching of rainbow trout embryos is reported to have been stimulated by treatment with nitrogen gas. In the pike *Esox lucius*, it was also reported that a reduced oxygen concentration in the culture medium (lower than 3.4 ppm) accelerated hatching. On the contrary, the hatching of *Fundulus* embryos is markedly retarded by a fully oxygenated medium, but boiling of the medium removes the inhibitory activity. Elevated oxygen concentration has been found to retard the hatching of pike and of the bream *Abramis brama L.* This line of study on *Fundulus* hatching has been carried out by *DiMichele* and *Taylor* (1980), who found that embryos incubated in water with dissolved oxygen concentrations greater than 6 ml O_2/1 delayed hatching indefinitely, while they hatched normally when placed in water of 4 ml O_2/1 or less. Moreover, they reported that incubation of *Fundulus* embryos in air resulted in a marked retardation of hatching. The air-incubated (hatching-retarded) embryos hatch within a short period of time after their reimmersion in water. A similar result was observed also in medaka embryos. Such a curious phenomenon was first documented by *Stockard* (1907; *Atz*, 1986). Air incubation would provide the embryos with a high pressure of oxygen, which, like a highly oxygenated medium, results in retardation of hatching enzyme secretion. Effects of the *"air incubation"* on hatching are of some interest also from an ecological or ethological viewpoint and will be discussed again in the following section. An apparently strange observation that potassium cyanide in low concentrations accelerates hatching in medaka can be reasonably understood in that this reagent causes hypoxia or anoxia in an embryo by affecting its respiratory activity. *DiMichele* and *Powers* (1982) found that the hatching time was different between two lactate dehydrogenase-B (LDH-B) genotypes of *Fundulus heteroclitus*. It seemed that the difference was related to the difference in their developmental rate. In the course of their studies, however, *DiMichele* and *Powers* (1984a) have proposed that hatching is stimulated when a growing respiratory demand of the embryo creates a hypoxic condition in the microenvironment surrounding the egg.

2. Temperature

Temperature is an important factor influencing the hatching enzyme secretion. In the case of the vendace *Coregonus albula*, temperatures as low as 1-2°C not only delay hatching markedly but also reduce the survival rate to as low as 6.5-47%, while a temporary exposure of the cooled embryos to higher temperatures, such as 8-12°C, accelerates their hatching and increases the extent of hatching to 82-96%. Thus, a previous cooling of the coregonid embryos, followed by an elevation of the water temperature, facilitates synchronization of hatching and controls the hatching time. Such treatment seems to be of great practical value in culture of this fish. A similar observation was also reported in the related lake whitefish *Coregonus clupeaformis*. In this case, it seems that a rise in temperature stimulates enzyme secretion. Once the enzyme is secreted, it will solubilize the egg envelope faster at higher temperatures than at lower temperatures, bringing about earlier hatching.

3. Light

Light is also one of the important environment factors that may influence fish hatching. In studies of dopaminergic regulation of fish hatching, *Schoots et al.* (1983b) reported that in medaka and zebrafish embryos cultured in a light-dark (12 h light/12 h dark) cycle, the hatching rate was significantly higher in the light period than in the dark period. In our recent experiments, development and hatching of the medaka embryos cultured were compared under conditions of constant light, constant darkness, or 14 h light/10 h dark (14L/10D); the results showed that the developmental rate from fertilization to hatching was not affected by any of these conditions, but hatching was significantly suppressed under the constant-dark condition. Moreover, if the embryos were allowed to develop before hatching in a 14L/10D cycle, a 14L/10D rhythm was observed in their hatching pattern. Therefore, hatching-enzyme secretion seems to be controlled by stimulation of photoreceptors such as eyes (and/or pineal gland?), probably via the central nervous system.

Ecological Factors Controlling Fish Hatching

In all animals, actual emergence of embryos from their envelopes takes a negligibly short time in comparison with their whole life history. However, hatching is a crucial event in their lives, and when and under what conditions hatching occurs have a great influence on their posthatching life.

As stated already temperature, oxygen supply and light are environmental factors influencing fish hatching in nature.

Among these, oxygen supply is a factor whose influence on hatching of some fish is especially interesting from ecological and ethological viewpoints. *Harrington* (1959) described delayed hatching in *fundulus confluentus* embryos that were not immersed in water but were stranded in air on the moist leaves of some littoral plants. As mentioned before, *Stockard* (1907) found experimentally that *Fundulus* embryos could not hatch when cultured in a moist atmosphere. Later, *Taylor et al.* (1977) reported a similar phenomenon in a related killifish species. *Fundulus heteroclitus.* According to their observations, tidal fluctuations of a marsh that is the habitat of the killifish are 1-15m. Major peaks of spawning activity of this fish occur in conjunction with the night high tide for several days at the new or full moon spring tides. Many eggs were found to be laid on the inner surface of the primary leaves of a marsh plant, *Spartina arterniflora*, at a level that is exposed at low tide. The embryos stranded in air on the plant leaves develop at the same rate as those in water up to the time of hatching. However, the embryos held in air past the normal hatching time continue to develop at a reduced rate without hatching. These *"latent embryos"* begin to hatch within a very short time when they are placed back into water. Thus, it seems that the air-stranded embryos in the field remain unhatched until they are resubmerged by a subsequent high tide. These authors explain that reproduction of this type may have important benefits for survival of eggs, in addition to a reduced exposure to predators. Laying the eggs in plants high on the marsh eliminates the probability of their being dispersed to inhospitable habitats or covered by silt in the strong tidal currents, and the larvae probably enjoy an improved chance for survival by hatching and spending their first days in the protected pools at the base of plants. *DiMichele* and *Taylor* (1980) extended their experimental studies on the *"delayed hatching"* in connection with the mechanism of hatching-enzyme secretion and proposed that both water and low dissolved oxygen concentration are necessary for hatching. Similar *"retarded hatching"* in stranded embryos can be observed also in the medaka, *Oryzias latipes*, although no stranding of embryos occurs in the normal life history of this species. When medaka embryos were kept on a moist filter paper, being partially dehydrated but receiving a fully adequate oxygen supply, they showed a marked retardation of hatching. Electron-microscopic examination revealed that the hatching gland cells of the hatching-retarded embryos seemed to be fully matured but showed no sign of initiation of secretion. The embryos in *"retarded hatching"* were found to hatch within a short period of time after reimmersion in water. Thus, it seems that the retardation (or arrest) of hatching would be useful for synchronous hatching. Synchronization of hatching seems to be of special significance also in hatchery culture.

Various developmental arrests occur in annual fish embryos at different developmental stages. This seems to be a survival strategy, generating several subpopulations that will develop according to different schedules under varied environmental conditions. Among these arrests, there is a diapause of prehatching embryos (diapause III), which may have resulted from the intensification of *"delayed hatching"* or *"retarded hatching."* It is said that a short-term arrest phenomenon, "retarded hatching", is sometimes encountered among nonannual aphyosemions and other nonannual cyprinodonts.

A famous example of *"retarded hatching"* of the stranded embryos associated with semilunar rhythm of spawning is the grunion *Leuresthes tenius*. It is known that grunion have a spawning period from March to September, with successive spawning runs at the interval of approximately half a month. Spawning is performed at night when the fish come out of the water onto the beach to bury and fertilize their eggs in the sand. Highest tides occur with the full and new moon, and the tides are higher at night than during the daytime in the spawning season of grunion on the California coast. The spawning occurs during a receding tide series, when the high water levels are lessening each night. From 1 to 4 nights are utilized for spawning. The embryos buried in the wet sand develop, but the well-developed embryos do not hatch until they are reimmersed by a following new series of high tides. Similar spawning runs on the beach sand are also reported in some other fish such as *Hubbsiella sardina, Galaxias attenuatus,* and *Enchelyopus cimbrus,* and the puffer *Fugu niphobles.* In the last case, however, it is not necessarily accepted that the spawned eggs are stranded alive in moist sand and hatching of the embryos is retarded until they are reimmersed in a following high tide. Thus, *"retarded hatching"* associated with the egg stranding may be of ecological significance in some special groups of fish such as cyprinodonts and some other related fish groups.

8

Hormonal Control

Most animals are seasonal breeds while few of them are polybreedus. In most animal species breeding occurs during restricted periods of the life history. The endocrine system provides a relatively slow acting link between the external environment and the internal state. The result is that reproductive behaviour is nicely synchronized with the maturation of the gonads and the environmental conditions appropriate to breeding. The term *"reproductive behaviour"* refers a divers range of activities involved in a propagative function including sexual, parental, and nestbuilding behaviours. Agonistic and migratory behaviour are also considered "reproductive" insofar as they are essential preliminaries to, or are involved in reproduction.

It is well established that in all classes of vertebrates the pituitary gonadotropins and gonadal steroids are the hormones most directly involved in the regulation of reproductive behaviour. There is a number of ways in which the CNS-pituitary-gonad axis may regulate reproductive behaviour; pituitary hormones may act directly on behaviour control mechanisms; pituitary hormones may stimulate the secretion of gonadal hormones which in turn regulate behaviour activity; both pituitary and gonadal hormones may act synergistically to regulate behaviour. In addition, these hormones may influence behaviour in fish. Other hormones may affect reproductive behaviour indirectly by their role in the development of secondary sexual characteristics.

The CNS-pituitary-gonad axis is influenced by internal and external factors acting through the sense organs and brain. Studies involving birds and mammals, particularly those of *Lehrman* (1965) and *Hinde* (1965) with ring doves and canaries, respectively, have emphasized that the integration and coordination of the components of reproductive behaviour depend upon a continuous interplay of external and internal factors; not only do hormones affect behaviour but also behaviour affects the endocrine state. Although the hormones of the gonads and anterior pituitary seem to play the major role in controlling reproductive processes, there is some evidence that neurohypophysial and thyroid hormones are also directly involved in the regulation of certain components of reproductive behaviour indirectly by their effects on general metabolism and growth and will not be considered here.

In general, hormones serve as relatively slow acting chemical links between the environment and an effector system. Thus there may be long-term seasonal changes over weeks or months, as in the annual cycles of very many animals. Changes may take place over hours or days, e.g., the switch from sexual to parental phases in birds. In addition, there are some situations in which it appears as if hormones may be involved in short-term changes over periods of minutes or seconds, e.g., the spawning of the female in response to courtship in the medaka, *Oryzias latipes* or the post-copulatory *"after reaction"* in the rabbit.

There is a danger of assuming that all reproductive behaviour must be regulated by hormones. It is possible that only certain components of the behaviour repertoire are under hormonal control, while other activities which are usually associated with reproduction may be causally linked to hormone-regulated activities without being themselves under the influence of the hormone state. In rodents, initiation of maternal responsiveness may depend on the endocrine state associated with parturition, while its maintenance is dependent on a nonendocrine mechanism stimulated by the presence of pups.

Certain behaviour patterns associated with reproduction may occur outside the breeding season in other functional contexts. For example, agonistic behaviour may be associated with reproduction, but it may also occur at times and places unrelated to breeding. Such activities are unlikely to be under the same hormonal control as seasonally changing behaviour and may not be regulated by hormones at all. This emphasizes the need for care in the use of terms included under the general heading of reproductive behaviour. Clearly, it would be misleading to refer to a behaviour pattern obviously associated with reproduction during the breeding season as "reproductive" when it occurs out of the season in a different functional context.

Like most animal species fishes also posses a CNA-pituitary-gonad axis. Furthermore, a wide range of steroids has been identified from male and female fishes. Surprisingly, there have been few unequivocal demonstrations for any fish that reproductive behaviour is in fact induced or regulated by gonadal steroids. Many workers have demonstrated effects of exogenous steroids on sex determination and the development of secondary sexual characteristics in fishes has but in most cases there is little mention of the effects of these treatments on reproductive behaviour.

MALE GONADAL HORMONES

In fishes the reproductive behaviour and the annual cycle have been the subject of more intensive studies. Most of infamations regarding reproductive behaviour have been collected from the three-spaired stickleback. There are two distinct races of the stickleback: the *Gasterosteus leiurus* form which spends its entire life cycle in fresh-water, the *trachurus* form which breeds in fresh-water in spring and summer but migrates to the sea in the fall where it remains until a return migration to freshwater in the spring. There are a number of consistent morphological and ecological differences between the two races. *Hagen* (1967) found that the two races are almost completely reproductively isolated and questions their inclusion in a single species.

In spring the males, which were previously associated in loose bisexual schools, become spaced our and acquire territories which they defend against other males. At the same time males acquire a distinctive appearance: red on throat and belly, dark back, and a blue iris. Internally, the kidney tubules begin to secrete mucus which is used in gluing materials used in nestbuilding. From this stage the male may go through one or more reproductive cycles lasting 20-30 days. Each cycle involved a series of overlapping phases or subcycles: *nestbuilding, sexual* and *parental.* Throughout these cycles the male defends his territory against males and other intruders. *Baggerman* (1966) also traced cyclical changes in displacement fanning and frequency of comfort movements.

Castration of the male in full breeding condition brings about a reversion to the nonbreeding condition—loss of nuptial colouration, a reduction in kidney tubules and cessation of sexual and nestbuilding behaviour. However, the agoniste behaviour of Castrated fishes have been made shortly before nestbuilding had commenced; and held under long photoperiod (16L:SD). In contrast, agonistic behaviour decreased along with other reproductive behaviour in males placed under short photoperiod (8L : 16D) after castration.

Baggerman (1966), working with the *trachurus* form, carried out a more detailed study of the effects of castration at different stages of the reproductive cycle. Castration of animals in full breeding condition, 9-19 days after the onset of breeding, resulted in a decrease in the frequency of swimming movements and a reduction in agonistic, sexual, nestbuilding, and displacement fanning behaviours to the level exhibited by intact animals after the end of the breeding period. In this study the onset of breeding is taken as the day a male built its first nest of the season in the presence of a male separated by a glass partition. Males castrated about 1 week before they were expected to build their first nest showed a decline in swimming activity and sexual behaviour to normal postbreeding levels; nestbuilding and displacement fanning disappeared completely. In contrast, agonistic behaviour increased to just below the level achieved by males in breeding condition and remained high for at least 3-4 weeks.

Wootton (1968) suggests that it is important to maintain the distinction between the *leiurus* and *trachurus* forms of the three-spined stickleback. Working with *leiurus*, Wootton finds that males castrated in the prenestbuilding stage and held under long photoperiod show a high level of aggressive behaviour when presented stimulus males or models in a test situation. Fish castrated after they have built nests also maintain a level of aggressive behaviour, not less than that of pre-breeding fish but lower than control fish with nests present. This result contrasts with Baggerman's finding (1966) that in *trachurus* castration of fish which have built nests results in a marked reduction in aggressive behaviour. In spite of important differences in testing procedures, which make direct comparison difficult, it may be tentatively suggested that the two findings reflect real differences in the endocrine mechanisms underlying the reproductive cycle which presumably are related differences in the biology of the two forms.

These findings concerning the control of agonistic behaviour raise the semantic. Each worker has used different testing procedures which tend to emphasize different aspects of the agonistic response : *Hoar* (1962a,b tested small groups of males and females in a situation in which few males succeeded in acquiring territories and building nests: *Baggerman* (1966) and *Wootton* (1968), using different procedures, tested isolated males which in many cases were able to build nests. Quite clearly, it is important to distinguish between agonistic behaviour involved in difference of a nest and territory, and agonistic behaviour without reference to a territory. In the latter case agonistic behaviour may provide a mechanism of general dispersion and perhaps simply represent the reaction of a fish to an intruder within its *"individual distance."* Defense of a territory is undoubtedly an important component of reproductive behaviour, whereas defense of "individual distance" is a much more general phenomenon and cannot be considered as exclusively reproductive in function. Therefore, although the agonistic motor patterns involved in both contexts are similar, it is not surprising that the underlying causal mechanisms appear to be somewhat different.

The acquisition of nuptial colouration and the onset of breeding behaviour has been correlated with evidence of increased steroidogenesis. *Gottfried* and *van Mullen* 1967) identified *testosterone, androstenedione, dehydroepiandrosterone,* and *progesterone* in the testes of male sticklebacks. They estimated that *"dominant" males,* which had acquired territories, had 5-7 times the steroid level of fish caught at the same time but which had not been allowed to pass into full breeding condition.

Replacement therapy has confirmed the role of *androgens* in inducing secondary sexual characters and reproductive behaviour in the stickleback. Methyltesterone treatment stimulates kidney tubule development and male nuptial colouration in gonadectomized adult males and females as well as intact juveniles of both sexes. However, juvenile fish failed to perform any nestbuilding behaviour, and onnly 5 out of 86 gonadectomized females held under long photoperiod built nests. In these five fish most of the nestbuilding movements were present but occurred sporadically and resulted in poorly constructed nests. Masculinized females never showed sexual behaviour toward introduced females. In the case of castrated males treated with methyltestosterone, *Hoar* (1962a) found that in those maintained under long photoperiod (16L : 8D) 87.5% built nests, whereas only 57% of males under short photoperiod (8L:16D) built nests and in addition took more than twice as long to commence building. The apparent refractoriness of androgen-treated fish under short photoperiod leads Hoar to suggest that although reproductive behaviour requires the gonadal hormone its full expression occurs only when gonadotropic activity of the pituitary is maintained at a high level by long photoperiod. *Smith* and *Hoar* (1967) investigated the endocrine involvement in displacement and parental fanning behaviour. They found that low doses of mathyltestosterone restored only the earlier stages of reproductive behaviour in castrate males, i.e., digging, collecting and gluing. Pushing and displacement fanning appear only after treatment with higher doses of androgen. Castration early in the parental phase resulted in a decline in fanning even though live eggs were present in the nest. Smith and Hoar conclude the displacement and parental fanning are regulated by a testis hormone.

Baggerman considered the possibility that gonadal hormones might be involved in the prespawning migration into fresh-water of the *trachurus* form of stickleback. Using salinity preference as a measure of migration disposition, *Baggerman* found that gonadal maturity coincides with a change in preference from saltwater to fresh-water, whereas gonad regression was correlated with a change in preference from fresh-water to saltwater. But, since similar changes in salinity preference can be induced in gonadectomized fish by manipulation of the photoperiod and temperature, it is clear that gonadal hormones are not directly involved. *Baggerman* (1962) did, however, find the testosterone treatment resulted in an increase in locomotory activity of a type associated with migrating fish.

Poeciliids have been used in numerous studies in which attention has centered on the effects of steroid hormones on morphological secondary sexual characters and on sex determination. In a few cases the effects of these treatments on behaviour are mentioned. *Cohen* (1946) and *Tavolga* (1949) found that treatment with *pregnenolone* induced male characters in maturing, genetically female platyfish, *Xiphophorus* (= *Platypoecilus maculatus*. These sex-reversed *"males"* showed male courtship patterns and attempted to copulate but, according to *Tavolga* (1949), courted less vigorously than genetic males under the same treatment. In contrast, genetic males treated with estradiol benzoate are said to have behaved like females and were pursued by normal males.

Hildemann (1954) was able to induce male courtship behaviour in female platyfish by treatment with methyltestosterone. It has been observed that females of *X. helleri* treated with *testosterone propionate* rose in the pecking order until a reversal in sexual behaviour occurred. As the sword and gonopodium developed, the sex-reversed *"male"* made attempts to copulate. Treatment of females with *extradiol dipropionate* failed to induce any change in social rank. These workers found that gonadectomized males and females maintained rank position for several months. They concluded that male sex hormone brought about a rise in rank of the female only by reversal in sex.

With testosterone the treated immature and adult genetic females of *Xiphophorus (=Platypoecilus variatus* performed what is referred to as phase 1 courtship which is characterized by a lateral display. Even after several weeks of treatment during which gonopodia developed in the young fish, the genetic females failed to perform the zigzag dance and the mating attempt.

Evidently the movements which initiate a courtship sequence are similar to those involved in agonistic behaviour, and perhaps sex recognition depends upon the response of the female to an initial threat by the male. Thus, it appears that phase 1 activities should be regarded as agonistic behaviour common to both male and female. Under normal circumstances agonistic behaviour occurs more frequently among males than among females. Androgen treatment of genetic females increases the frequency of agonistic behaviour and in this respect has a masculinizing effect but is relatively ineffective in inducing movements which are functionally restricted to the sexual context.

Gonadectomy of the male platyfish, *Xiphophorus maculatus*, resulted in a decrease in frequency or duration of copulation thrusts, swings and sidling whereas pecks, backs, and approaches remained the same Clemens *et al.* (1966) treated guppies with *testosterone* from birth to 60 days. After treatment they found a very marked increase in the proportion of males—in some cases the males outnumbered females 9:1. However, when paired with females only 14% of these males, including sex-reversed genetic females, sired young, even though in many cases they showed full male colouration and yielded viable sperm on stripping. *Clemens et al.* suggest that the failure to breed resulted mainly from a behavioural deficiency which was not simply the result of a lack of endogenous androgen. A possibility, suggested by the present author's work with the guppy, is that males failed to respond to female because of the prolonged isolation from respective females.

Working with the medaka, *Oryzias latipes, Okada* and *Yamashita* (1944) confirmed by castration and testosterone treatment that the male secondary sexual characteristics are under hormonal control. Testosterone treatment of females or implantation of a testis results in masculinization and the performance of male behaviour including pursuit of normal females. *Yamamoto* has confirmed that complete functional sex reversal may be achieved with androgen treatment of genetic female medakas. Unfortunately, there has been no detailed comparison of the behaviour of normal males and sex-reversed genetic females.

Tavolga (1955) found that castration of male gobies, *Bathygobius soporator*, abolished aggressive behaviour toward an introduced male. Instead the operated males failed to discriminate between male males and females, and between gravid and nongravid females, and courted all equally. Spawning behaviour of castrated males with gravid females appeared to be normal. The male brooded the infertile eggs which resulted from the spawning. *Tavolga* (1956) also investigated the stimulus situation involved in prespawning behaviour of goby males. Intact males readily distinguished gravid females from nongravid females and males. However, if stimulus fish were presented in glass containers males failed to discriminate and courted all females and nonaggressive males equally. This suggests that normally the males recognize the gravid female on the basis of chemical cues. *Tavolga* (1956) demonstrated that a substance present in the ovary of gravid females, and presumably released through the genital pore, will elicit a vigorous courtship response in a male, even in the absence of a stimulus fish. *Tavolga* (1956) suggests that in the male goby a gonadal hormone affects the sensitivity of the olfactory organs perhaps by making the male differentially sensitive to chemical factors released by the gravid female. Castration may lower the sensitivity threshold and permit chemicals other than the ovarian fluid to stimulate courtship behaviour.

Noble and *Kumpf* (1936) reported that males of *Hemichromis bimaculatus* performed typical courtship, fertilization and brooding movements for 202 days after castration. The males developed nuptial coloration and genital tubes at each spawning. After a normal spawning both parents are inhibited from eating their young and perform brooding behaviour which consists of collecting and guarding the young. *Noble et al.* found that brooding in response to donated young could be induced by treating nonbrooding fish, both male and female, with a variety of hormones. Positive responses were only obtained with fish which had previous experience of brooding and/or spawning. Thirteen fish, including four castrates, began to brood normally after treatment with corpus luteum extracts (source not given presumably rich in progesterone). Prolution and prolactin were also highly effective in inducing brooding. Anterior pituitary extract, fresh fish pituitary, thyroxin, desiccated thyroid, and testosterone were also effective in a smaller proportion of fish treated. Nearly half the fish given control phenol injection also began brooding.

These results are difficult to interpret but serve as a useful warning of (a) the role of experience in determining the outcome of hormone therapy, and (b) the possibility that a wide variety of agents may produce the same behavioural effects: In this case it seems likely that some or all the agents were exerting their effects indirectly, perhaps by stimulating pituitary secretory activity.

Aronson (1951) found that castration of male *Tilapia macrocephala* resulted in a reduction of the genital tubes and loss of the yellow colouration of the operculum. Castrated males continued to dig nests as frequently as intact males. The effects of castration were reversed by testosterone treatment Previously, *Aronson* and *Holz-Tucker* (1947) had shown that testosterone treatment of ovariectomized and intact females resulted in the growth of the genital papilla and the acquisition of male colouration on the operculum. Working with another species, *Tilapia mossambica, Clemens* and *Inslee* (1968) obtained functional sex reversal in genetic females by treating them with methyltestosterone for the first 69 days of life. The sex-reversed fish exhibited male mating colouration and nestbuilding behaviour when placed with ripe females. *Aronson* (1959) compared the reproductive behaviour of three males of *Aequidens latifrons* before and after gonadectomy. Up to 6 weeks after castration all elements in the mating pattern were still present, most of these showing little change in frequency of occurrence or duration from their own preoperative levels. There was an increase in nest passing and several other items of behaviour after castration. The increase was particularly marked for the period 1 day after spawning. On the other hand, there was a noticeable decline in nestdigging after castration.

As respected by *Noble* and *Kumpf* the gonadectomized males of the anabantid, *Betta splendens,* perform courtship movements, whereas gonadectomized females do not. These authors also found that a small number of ovariectomized *B. splendens* females developed a testis and in such cases acquired male colouration and behaviour. Also working with anabantids, *Forselius* (1957) reports that sterile *Colisa labiosa xlalia* hybrids acquired male secondary sex characteristics and exhibited "migratory" and reproductive behaviour. Histological examination of a large number of the hybrids failed to reveal any trace of gonadal tissue. However, evidence that androgens are involved in the control of male characteristics is provided by the finding that testosterone propionate treatment of females of *C. lalia* induced male colouration and, in some cases nestbuilding and defense.

Johns et al. (1969) found that some males of the blue gourami, *Trichogaster trichopterus*, performed reproductive behaviour after castration. Of a total of 16 males castrated, 11 failed to build bubble nests or show any sexual behaviour. The other five built nests and *"spawned"* within 7 days of being paired with female. Apart from a relatively long delay between introduction of females and actual spawning, the reproductive behaviour of these five did not differ from that of control fish. Castrate males which had spawned readily accepted fertile eggs and performed parental behaviour for several days. Fertile eggs given to nonspawning castrates were eaten. Castrate males placed with intact or other castrate males performed agonistic behaviour until a dominance relationship was established. Agonistic behaviour of castrates did not obviously differ qualitatively or quantitatively from that of intact males. The dorsal fin of the male gourami is longer and more pointed than that of the female. The dorsal fin of castrated males became shorter and rounded, more like that of the female. Significantly, however, the dorsal fins of the castrates which spawned were more like those of intact males than were the fins of nonspawning castrates. Castrated males treated with methyltestosterone developed the long pointed dorsal fin and in all cases spawned when paired with ripe females. These results indicate that sexual behaviour and the characteristic male dorsal fin are under gonadal hormone control. Of the five castrates showed identifiable regenerating testis.

Machemer and *Fiedler* (1965) investigated the hormonal involvement in nestbuilding of another anabantid, the paradise fish, *Macropodus opercularis*. Two out of eight intact males given only *methyltestosterone* showed an increase in building tendency but, apparently because of inadequate mucus production, building was incomplete. *Testosterone* in combination with prolactin resulted in nestbuilding in three males and two females. *Machemer* and *Fiedler* do not appear to have run a control series, and it is important to note that, in at least some species of anabantid, isolated males begin nestbuilding apparently spontaneously.

Greenberg and *Hale* note that in the green sunfish, *Lepomis cyanellus*, immature and mature fish of both sexes perform agonistic behaviour. Although defense of a nest is limited to sexually mature males. There was no decrease in agonistic behaviour after gonadectomy in a small number of males and females placed in a test situation. *Smith* (1967) found that castrated males of *Lepomis megalotis* and *L. gibbosus* maintained high levels of aggressive behaviour but failed to build nests. Treatment with testosterone restored nestbuilding but, in the case of *L. megalotis* placed in large experimental pools, the males did not cluster together to the same extent as sham-operated fishes.

In the male of the viviparous sea-perch, *Cymatogaster aggregata*, the occurrence of full reproductive behaviour coincides with maximum gonadal development and histochemically demonstrated steroidogenesis. The gonads of fish in winter condition (October) matured in fish placed under long photoperiod (16L:8D) and high temperature (20°C). Such fish also acquired the secondary sexual structures and performed sexual behaviour including chasing, heading-off the female, courtship dance, lateral quiver, and mating attempt. Castrated fish under the same conditions did not acquire the sexual structures. However, they still performed some of the elements of sexual behaviour: chasing and mating attempt, but without the darkening characteristic of full reproductive behaviour.

Castrated males treated with methyltestosterone acquired the secondary sexual characteristics and showed an increase in sexual behaviour, although the level of sexual behaviour did not reach that of intact males under the same environmental conditions. Significantly, chasing and mating attempts which persisted after castration occur in intact fish with reduced vigour and frequency throughout the year. Thus, it appears that certain components of the sexual behaviour occur only at times at which insemination is known to occur. These activities disappear after gonadectomy and reappear as a result of testosterone therapy. In contrast, chasing and mating attempt appear to be to some extent independent of gonadal control and for much of the year are not obviously associated with a reproductive function.

Little is known of the role of gonadal hormones in the reproductive behaviour of salmonids. *Jones* and *King* (1952) castrated four adult males of *Salmo salar*. In three castrated males treated with testosterone propionate sexual behaviour was partially restored: males followed females and one male quivered. The one castrate which was not treated with androgen showed no interest at all in the females. In all species of fishes studied, male secondary sexual characteristics and at least some components of reproductive behaviour are under gonadal hormone control. In a number of cases it has been claimed that complete reproductive behaviour persists in castrated fish. The most striking example of this is provided by *Tavolga's* study (1955) of *Bathygobius soporator*. Others involve cichlids and anabantids. However, *Johns et al.,* have reason to suggest that in the gourami the persistence of reproductive behaviour may result from the presence of unidentified regenerated testis or the existence of an extragonadal source of *androgen*.

FEMALE GONADAL HORMONES

Comparatively less attention has been paid to hormonal regulation of female behaviour than that of the male. This is at least partly because the female is in most cases a more passive partner. Even where nestbuilding and parental care occur, in many species it is the male which is responsible. Thus, the behaviour of the female is usually difficult to detect or quantity, with the result that it may be difficult to assess the effects of gonadectomy or hormone treatment.

Numerous investigators have applied *estrogen* treatment to male and female poeciliids. Indeed most investigations have been concerned with the effects of *estrogens* on the gonads and morphological structures. In general, *estrogens* suppress the testis and male characteristics in developing genetic males. When treatment is withdrawn the testis and male gonopodium and, colouration develop. None of these investigations have given any clear indication that estrogenlike hormones are inovolved in the regulation of female behaviour.

Ovariectomy of poeciliid females has been performed in a small number of investigations. It is found that females of *Xiphophorous helleri* and *Poecilia (=Mollinesia)* sp., respectively, remained sexually attractive to males after ovariectomy. However, there is no indication that an attempt was made in either of these studies to assess female responsiveness to male courtship. It is well-known that poeciliid males will direct persistent and vigorous courtship toward unresponsive females.

Liley (1966) has described the sexual response of the female guppy, *Poecilia reticulata*, and finds that nonvirgin females go through a cycle of receptivity which can be correlated with the cycle of brood production. Females respond most readily and are more likely to accept copulation in the few days following the birth of a brood of young. Virgin females ar more persistently receptive until fertilization of the eggs occurs. Ovariectomy of virgin females result in an initial decline in receptivity over a period of 5-6 days. However, nonvirgin females tested several weeks after gonadectomy still occasionally responded to male courtship. It was concluded that sexual behaviour may still occur in the absence of gonads, but that in intact fish the ovary exerts a regulatory and perhaps stimulatory effect upon a more direct pituitary or neural control. The importance of the gonadal contribution is emphasized by the finding that treatment of hypophysectomized females with the gonadotropin fraction of salmon pituitary extract brought about a restoration of sexual responsiveness only after the gonad recovered from its regressed state resulting from hypophysectomy.

In the medaka, *Oryzias latipes*, it is found that ovariectomy of females and estriol treatment of adult males had no effect on secondary sexual characteristics. However, *Yamamoto* (1962), and *Yamamoto* and *Matsuda* (1983) have established that functional sex reversal of genetic males may be induced by treatment of young with estrogens. Sex-reversed genetic males are fully functional as females and breed with true males. Unfortunately, there has not been a detailed behavioural comparison of sex-reversed genetic males with normally developing females.

Little is known of the role of gonadal hormones in the sexual response of the medaka. *Egami* (1955) found that treatment of females with estradiol benzoate brought about a reduction in the occurrence of oviposition. This result may have been caused by the suppression of pituitary gonadotropin secretion and the consequent regression of the ovaries.

In one species *leiurus* form, aggressive behaviour was measured by placing two fish of the same sex in a tank and scoring the number of attacks during a series of 5 min observations over a period of several weeks. The results, confirm earlier findings that prenestbuilding agonistic behaviour of males is hardly affected by castration. Intact females attacked each other at about one-third of the rate shown by intact males. Ovariectomized females performed twice as many attacks as intact females. These results suggest that ovarian secretions normally suppress aggressive behaviour to some extent. *Wai* and *Hoar* noted that most of the intact females which showed high levels of aggressive behaviour were immature and therefore, presumably, estrogen levels were low. The fact that gonadectomized females were less aggressive than gonadectomized males suggests that there are genetic sex differences in the levels of aggression, or that the presence of gonadal hormones earlier had some persistent effects. *Forselius* (1957) found evidence that estrone injections resulted in a marked reduction in aggressive behaviour in males of *Colisa Lalia*. In contrast to their findings with castrated males, *Noble* and *Kumpf* (1936) found that Ovariectomy of females of *Hemichromis bimaculatus* resulted in total loss or reproductive behaviour. Injection of ovarian extract restored most of the sexual behaviour. It is well proved that brooding behaviour could be induced in experienced females by treatment with a variety of gonadal and pituitary hormones. Of these agents corpus luteum extract proved to be one of the most effective. Ovariectomy of *Tilapia macrocephala* brought about a reduction of the genital tube and the reappearance of a silvery immature condition of the operculum. *Estradiol* treatment of ovariectomized fish induced the growth of the genital tube but had no effect on the silvery operculum. *Aronson* found (1951) that the majority of intact females built nests prior to spawning. Ovariectomy resulted in a reduction in nestbuilding to about the level of intact or castrate males, that is, nests were build in about 11% of the pairings. *Aronson* notes (1957) that in *Tilapia macrocephala* completion of ovulation is marked by the occurrence of nest-passing, an activity closely associated with spawning.

Normally the female of blue gourami initiates the spawning sequence by approaching and butting a dark, nestbuilding male. None of the eight ovariectomized females made sexual responses when exposed to males in reproductive condition. Two ovariectomized females treated with estrone also failed to respond. In addition, ovariectomized females were evidently less attractive to males than were intact females: Males directed significantly fewer contacts with their long pelvic rays toward ovariectomized females. There is histological and behavioural evidence that the pelvic rays are chemosensitive. Thus, the effect of ovariectomy on female *"Attractiveness"* suggests that the ovary is responsible directly or indirectly for the elaboration of a specific chemical detected at the female body surface by the male. Such a secretion might provide the means by which males discriminate between males and females and between gravid and nongravid females. It should be noted however that *Picciolo* (1964) was unable to find behavioural evidence that male anabantids responded to chemical stimuli during reproductive behaviour, although he did find that amputation or cauterization of the pelvic fins reduced reproductive success.

Ovarian involvement in the production of a chemical attractant has been well-established by *Tavolga* (1956) in his work with *Bathygobius soporator*. The chemical has not been identified but is evidently only produced in effective quantities in gravid fish ready to spawn. *Amouriq* (1964, 1965a, b) has concluded that a substance produced in the ovaries of female guppies which elicits increased male activity, is in fact an *estrogen. Ball* (1960) reviewed much of the work concerning the hormonal control of ovipositor growth in the bitterling, *Rhodeus amarus.* Early work had suggested that ovipositor growth is stimulated by ovarian hormones. However, *Brestchneider* and *de Wit* (1947) and other found that ovipositor growth could be induced by a wide variety of chemical and physical stimuli. Ball suggests that although there is no reason to suppose the normal ovipositor growth is not under ovarian steroid control, in many experimental situations it was mainly a response to internal changes involved in a stress reaction. *Shirai* (1962, 1964) finds a clear correlation between ovipositor growth and ovarian condition in the Japanese bitterling, *R. ocellatus.* He notes that there is a period of slow steady growth in the prebreeding season and a periodic fluctuation in length during the breeding season. *Shirai* (1964) speculates that there may be two hormone factors involved: an estrogenlike hormone, which stimulates the long-term growth, and another factor involved in the short-term cyclical changes during the breeding season. In this last phase fish show maximum length of ovipositor at the time that ripe eggs are extruded into the ovarian lumen, and it is at this time that oviposition occurs. If there is no mussel present, maturation, ovulation, and ovipositor growth are all inhibited. The clear correlation between ovipositor growth, ovulation, and oviposition suggests that gonadal hormones are in some way involved in the regulation of the spawning act. *Shirai* (1962) makes an alternative suggestion that a physical stimulus caused by the extrusion of mature eggs into the lumen brings ovipositor growth under nervous control.

Apart from an obvious correlation between ovarian maturation and spawning readiness, there is no direct evidence that gonadal hormones are involved in the regulation of spawning behaviour in female fish. There are relatively few reports of the experimental use of steroid estrogen, none of these has revealed behavioural responses to hormone treatment. *Sundararaj* and *Goswami* (1966) treated sexually mature but unovulated female catfish, *Heteropneustes fossilis*, with estradiol benzoate, testosterone propionate, and progesterone 3 days after hypophysectomy. These steroids failed to induce ovulation or spawning behaviour. In conntrast, three corticosteroids induced ovulation and, in some cases, oviposition in the absence of males. Thus, it appears that corticosteriods can act as ovulating agents and proved almost as effective in this regard as mammalian luteinizing hormone (LH). Furthermore, hypophysectomy of the experimental fish ruled out the possibility that the steroids exerted their effect by an action involving the pituitary. *Sundararaj* and *Goswami* (1966) direct attention to investigations which have shown that in the Pacific salmon the *17-hydoxycorticosterone* titer in the blood progressively increases in fish as they migrate from the sea to spawning grounds. They speculate that perhaps the corticosteroids are directly involved in ovulation and spawning in fish. Changes in conticosteroid levels in migrating and spawning fish have usually been interpreted, e.g., *Chester Jones* and *Phillips* (1960) and *Fagerlund* (1967), as a response to activity and stress. However, considering the treatment with gonadal sterods has been so conspicuously unsccessful in inducing reproductive behaviour in female fish it seems worthwhile considering the possibility that corticosteroids may be in some way involved.

NONSPECIFIC EFFECTS OF GONADAL AND THYROID HORMONES

There are a number of indications that gonadal and thyroid hormones may have more general effects on behavioural activity. *Stanley* and *Tescher* (1931) reported a considerable increase (400%) in the locomotory activity of goldfish fed on ground mammalian testicular substance. This effect became marked within hours of feeding and continued for at least 24 hrs. *Hoar et al.* (1955) found that goldfish immersed in solutions of *thyroxine, testosterone,* or stilbestrol showed greater locomotory activity and were more responsive to electrical stimulation than untreated fish. *Forselius* reports (1957), without giving details. "An increase, of a appetitive behaviour and general locomotory activity" in the anabantid, *colisa lalia,* after injection of thyroxine, testosterone, and chorionic gonadotropin. *Thyroxine* treatment resulted in increased locomotory activity and jumping in guppies, *Poecilia reticulata. Hoar et al.* (1952) showed that chum salmon fry treated with methyl-testosterone and synthetic thyroxine became more active and showed less marked schooling than control or thiourea-treated fry. Coho and sockeye salmon yearlings treated with thyroxine, testosterone, or stilbestrol, and yearling sockeye treated with estrogens, showed a decrease in the time required to make a standard response to flowing water. *Van Iersel* (1953) found that male three-spined sticklebacks (*trachurus* form) treated in midwinter with *testosterone propionate* began *"fluttering"* (persistent swimming against the glass walls of the aquarium) which did not occur in control fish at that time. Iersel regards this fluttering as an expression of migratory behaviour. Both androgen and thyroid stimulating hormone (TSH) treatments resulted in significant increase in swimming movements and bouts of fluttering in castrated male stickle-backs. *Baggerman* suggests that TSH produced this effect indirectly by stimulating the secretion of thyroid hormones.

A different mechanism of hormone action is suggested by *Amouriq* (1964, 1965a,b) who found that male guppies, *Poecilia* (= *Lebistes*) *reticulata*, placed in water which had previously held females, exhibited a marked increase in their locomotory activity; water which had previously held males was ineffective. *Amouriz* added extracts of skin, intestine, and ovary of females to the aquarium water and found that only the latter induced a significant increase in male activity. An estrogen, hexestrol dipropionate, when added to the water, resulted in a marked increase in male activity. More recently *Amouriq* (1967) has established that the range of concentrations of the steroid affecting locomotory activity is quite narrow: 0.025 mg/ml evokes hyperactivity; 0.05 mg/ml results in hypoactivity; concentrations below the first and above the second have no effect. *Amouriq* (1965b) concludes that an ovarian hormone induces hyperactivity in males and maintains female attractiveness to males. Thus, he is suggesting that the hormone act as a pheromone in addition, presumably, to its endocrine role in the female.

Earlier studies, have emphasized the role of visual stimuli in guppy courtship. Although not tested experimentally there has been little evidence to indicate that chemical stimuli are involved. On the basis of data provided it is difficult to appreciate the functional significance of the response described by Amouriq. The response was only obtained using water from a 2-liter container which had held as many as 10 females for 24 hrs. Maximum locomotory effects occurred 5 hrs after the start of the addition of ovarian extract. The effects of hexestrol remained slight for 6-7 hr from the start of treatment after which there was a marked increase in male activity. This suggests that the hormone may not be acting directly as an external stimulus but is being taken up through the skin and gills of the male and acting on the central nervous system or sense organs, perhaps increasing the susceptibility of the fish to external stimuli and thereby bringing about an increase in locomotory activity.

Because of the delay in obtaining a locomotory response, the relatively high density of females required to obtain effective concentrations, the mobility of both male and female, and the lack of obvious contact behaviour in courtship it is difficult to see how the chemical released by the female could play an important role as a sexual attractant or even by simply increasing the searching behaviour of the male male. It is of coursse possible that more refined behavioural measures may reveal female-oriented responses at lower concentrations and with less delay.

This work with several species of fish indicates that hormones can affect behaviour by bringing about an increase in general activity. However, it is not clear whether the hormones influence activity by an-effect on general metabolism or whether they act directly on neural mechanisms underlying the behavioural responses. Evidence the steroid and thyroid hormones may directly affect CNS and sense organ states is provided by a number of electrophysiological studies. Several investigations involving goldfish have revealed that various steroids and thyroid hormones exert differential facilitatory and inhibitory effects on the excitability of the olfactory bulb and regions of the fore and midbrains.

Although little is known of behavioural significance of most of the above neurophysiological findings, they appear to indicate that steroid and thyroid hormones influence the responsiveness of the CNS to external stimuli. It seems reasonable to suppose that hormone-induced changes in responsiveness to general or specific external stimuli are in part responsible for increased locomotory activity. Such changes in activity and responsiveness may be important components in migratory and other reproductive behaviour in fishes.

THYROID HORMONES AND MIGRATORY BEHAVIOUR

Baggerman (1962) suggests that in the stickleback the gonads gonadal hormones are involved in timing and perhaps augmenting the action of the thyroid hormone which, she believes, plays an important causative role in the onset of migration in certain anadromous fishes. *Baggerman* tested the effects of *thyroxine*.TSH and various thyroid blocking agents on the salinity preference of the *trachurus* form of the three-spined stickleback. The results suggest that an increase in the level of circulating thyroid hormone induces a preference for fresh-water. This agrees with the fact that the stickleback migrates to fresh-water in spring when the thyroid gland shows signs of heightened activity. Low levels of thyroid hormone, as is believed to occur in the fall, result in a preference for saltwater. As mentioned previously, *Baggerman* (1962) also found that both thyroid hormone and androgen treatments stimulate a form of locomotory activity characteristic of migrating fish. *Baggerman* (1962) concludes that an increase in thyroid activity brought about by long photoperiod induces two changes which are known to be associated with migration. *Baggerman* (1960, 1963) carried out similar experiments with juvenile Pacific salmon, *Oncorhynchus*, and found an increased preference for saltwater associated with the downstream migration. The saltwater preference is in some species correlated with hightened thyroid activity. Juvenile coho salmon treated with TSH showed a change from fresh-water to saltwater preference. Thiourea treatment induced a fresh-water preference in coho which initially showed a preference for saltwater *Baggerman*, 1960). Similarly, underyearling pink salmon treated with thyroid blocking agents showed a change from saltwater to fresh-water preference. Although the changes in salinity preference induced by thyroid hormone are directly opposite in sticklebacks and salmonids. *Baggerman* (1962) argues that the thyroid gland is directly involved in the causation of migration of these species.

PITUTTARY HORMONES

A. Prespawning Behaviour

A renew of literature states that at least one pituitary hormone regulates gonadal development and hormone secretion and thereby exerts an indirect control over reproductive activities. Indeed, treatment of intact fish with fish or mammal pititary material has proved a highly effective approach to the breeding of commercially important fresh-water fishes. In a number of cases pituitary hormones may have a more direct influence on behaviour, perhaps by acting on the CNS directly. Much of the evidence for such a pituitary control is of an indirect nature. Usually the first indication that the pituary may have a direct effect on behaviour arises from experiments in which it is found that reproductive behaviour persists after gonadectomy. Of course, it does not necessarily follow in such cases that the pituitary is directly involved; other endocrine glands may be important, for example, the thyroid in stickleback migration, or there may be no endocrine involvement at all.

The work of *Baggerman* (1966) and *Hoar* (1962a,b) leaves little doubt that in the stickleback gonadal development and hormone secretion are under pituitary gonadotropic control. But, in addition to the indirect effects of the pituitary on reproductive behaviour, both of these workers have hypothesized that prenestbuilding aggressive behaviour is directly controlled by the pituitary. *Baggerman* (1966) found that castration of males prior to nestbuilding did not affect aggressive behaviour, whereas castration after the onset of breeding resulted in a marked decline in aggressive behaviour. Hoar's evidence (1962a,b) for a direct pituitary action is based upon a comparison of castrated sticklebacks under regimes of long and short photoperiod. The long photoperiods are assumed to produce a high level of pituitary gonadotropin, while the output of these pituitary factors is greatly depressed or eliminated under short photoperiods. Males castrated prior to nestbuilding show a high level of aggression if they are maintained underlong photoperiod. Castrates held under short photoperiod are much less aggressive. In experiments designed to measure the effects of a series of mammalian pituitary hormones on aggressive behaviour only treatment with LH consistently produced an increase in aggressive behaviour. *Ahsan* and *Hoar* (1963) also found LH to be the most effective mammalian gonadotropic hormone in eliciting gonadal development in immature fishes.

There are indications that in the stickleback *androgens* and *pituitary* hormones act synergistically to maintain full reproductive behaviour. *Hoar* (1962a,b,) found that methyltestosterone treatment was less successful in inducing nestbuilding in castrate males kept under short photoperiod than in males under long photoperiod. These findings suggest that although the behaviour of the nestbuilding phase requires the gonadal hormone its full expression only occurs when the gonadotropic activity of the pituitary is high.

Smith (1967) found that in males of two species of *Lepomis* aggressive behaviour remained high after castration and was not affected by androgen treatment. Aggressive behaviour remained high under short photoperiod provided the temperature remained high (25°C), but it declined at low temperatures (13°C), Treatment with human chorionic gonadotropin, although effective in stimulating nestbuilding, did not affect aggressive behaviour. Smith suggests that in these species aggression, although inhibited by low temperature, seems more closely related to social conditions than hormonal state.

Wiebe (1967) was able to relate the reproductive cycle of *Cymatograster aggregata* to the seasonal light and temperature cycle, indicating that pituitary gonadotropins are involved in stimulating gonard development and hormone secretion. Treatment with methallibure (ICI 33, 828), a substance shown to have antigonadotropic activity in a number of fishes species, effectively blocks gonadal development and results in the complete elimination of reproductive behaviour including chasing and mating attempt. As these two activities persist after castration *Wiebe* concludes that they are normally under direct pituitary control.

Burger (1941) hypophysectomized males of *fundulus heteroclitus* and noted that the later stages of spermatogenesis came to a halt. Implantation of *Fundulus* pituitary material, five pituitaries at each treatment, at 3-5 day intervals, resulted in vigorous sexual behaviour. By the tenth day the fishes were in a *Frenzy* of display." As the treatment also resulted in the recrudescence of the testes, it is not known whether the behaviour effects were the results of a direct action of the pituitary material on the CNS or a result of the stimulation of gonadal secretory activity. *Tavolga* (1955) hypophysectomized males of *Bathygobius soporator* and found that courtship and agonistic behaviour were eliminated. This result is evidently not simply an effect of a reduction in gonadal hormone secretion since castration does not eliminate courtship behaviour. Hypophysectomy of virgin female guppies, *Poecilia reticulata* results in an immediate and complete decline in sexual responsiveness. Because this effect contrasts with the relatively slow decline in sexual behaviour following gonadectomy, and because of the reappearance of sexual responses several weeks after gonadectomy, *Liley* proposed that there is a pituitary factor directly involved in the control of receptivity in the female. On the other hand, the decline in receptivity after gonadectomy and the cyclical changes in responsiveness of nonvirgin fish suggest that the ovary is also producing a hormone which regulates, or acts synergistically with, the more direct pituitary mechanism.

Liley and *Donaldson* (1969) treated female guppies with a partially purified gonadotropin fraction of salmon pituitary material. Several of a group of hypophysectionized, ovariectomized fish exhibited a low level of sexual responsiveness. Hypophysectomized females with their ovaries intact showed a marked increase in sexual responsiveness after several days of hormone treatment. This increases coincided with the growth of the ovaries which had previously regressed. These results appear to support the suggestion that in the female guppy sexual responsiveness is not completely dependent upon any one hormone, but under normal circumstances the gonadotropic and ovarian hormones act synergistically—the absence of one hormone may be partly compensated for by the other.

B. Spawning Behaviour

It is convenient to give separate consideration to the role of hormones involved in the actual spawning act involved in the release of gametes as opposed to the more lengthy prespawning procedures. There is some evidence that the two phases may be under separate controls. In general, the distinction between prespawning and spawning behaviour is more clearly defined in males than in females. In females there may be little in the way of overt prespawning behaviour before a female in "ripe" condition responds to male courtship and proceeds rapidly to actual spawning.

Pituitary hormone therapy has been used to accelerate maturity in fishes nonreproductive condition. More often these techniques have been applied to fish already mature to induce or accelerate *ovulation* and perhaps actual *spawning*. In most studies reported treated fish are stripped by hand and the eggs fertilized artificially. However, if stripping is not carried out but the females are placed with mature males, spawning occurs within a few hours of ovulation. *Fontenele* (1955) reports that in Brazilian fish culture practices a hormone-treated female is separated by a partition from a male until readiness to spawn is indicated by a state of accentuated agitation. The partition is then removed and the fish allowed to spawn. *Spawning* usually occurs within 24 hr of the first of a series of injections of fish pituitary material.

There is some evidence that in male fish readiness to spawn is associated with seminal thinning. This increases in fluidity of the semen appears to be the same as the process referred to as *"spermiation"* by *Yamazaki* and *Donaldson* (1968a,b). Both pairs of workers have demonstrated for a number of species of fishes that *spermiation* occurs in response to treatment with fish and mammalian gonadotropin preparations. More recently, *Yamazaki and Donaldson* (1968c) have found that androgen alone will induce spermiation in hypophysectomized goldfish. However, it is evident that spawning behaviour may occur, or be induced by androgen treatment, in castrated fish. Thus, spawning in males is not dependent upon the presence of intact gonads as appears to be the case in females.

At present there is little convincing evidence for or against the above suggestions regarding the significance of ovulation in the timing of oviposition. *Yamazaki* (1965) has gone further than most workers in an attempt to determine the factors involved in the regulation of spawning. Under natural spawning conditions the male goldfish courts the female for 1 to 2 days before ovulation and spawning occur. Evidently male courtship is important in the induction of ovulation. Spawning will not courtship is important in the induction of ovulation. Spawning will not occur in females from which ovulated eggs have been stripped, suggesting that the presence of ova is an important factor. On the other hand, females hypophysectomized after ovulation may oviposite normally about 3 hr after the operation. A small number of hypophysectomized females in which ovulation had been induced by hormone treatment 1—4 days after hypophysectomy spawned normally when paired with males. *Yamazaki* (1962)also found that male foldfish would spawn several hours, or in some cases, the day after hypophysectomy. *Yamazaki* (1965) concludes that, although ovulation is induced by pituitary gonadotropio hormone, actual spawning is not under direct pituitary control. However, he recognizes the possibility that in his experiments residual pituitary hormone may remain in effective quantities for some time after hypophysectomy. In addition he drawn attention to the possibility that neuro-hypophysial of hypophysectomized fish and could be involved in the regulation of spawning behaviour.

There have been a number of suggestions that neurohypophysial hormones are directly involved in the causation of spawning behaviour. *Wilhelmi et al.* (1955) found that purified fish and mammalian neurohypophysial preparations and synthetic oxytocin would induce the spawning reflex in *Fundulus heteroclitus.* The response usually appears within 10 min after injection and may persist for 20-30 min. This behaviour occurs in hypophysectomized or castrated fishes of either sex. However, there is no coordination of the response when males and females are injected and placed together. It is suggested that this response is mediated by direct excitation of a nervous center and that sex hormones play no part in mediating this phase of the sexual behaviour pattern.

Egami (1959) injected mammalian neurohypophysial substances into *Oryzias latipes* and observed spawning moments in both males and females isolated from males prior to treatment oviposited even though there was no male present. *Egami* and *Nambu* 1961) consider these findings in relation to the normal breeding activity of *Oryzias.* During the breeding season the female lays eggs almost every morning. Maturation of oocytes and ovulation may take place in a female which is isolated from a male, but oviposition only occurs naturally after a period of male courtship. *Egami* and *Nambu* suggest that stimuli arising from male courtship induce the secretion of neurohypophysial hormone which, when it reaches a certain level, causes oviposition to occur.

Females of *Gambusia affinis* performed behaviour resembling a spawning reflex and released their embryos shortly after treatment with oxytocin, or homogenates of from neurointermediate lobe or rat neural lobe. *Blum* (1988) injected reserpine into immature angelfish, *Pterophyllum scalare.* One to three hours after the injection the fish darkened; 6-8 hr after injection several of the fish performed typical spawning movements. *Blum* suggests that reserpine may exert its effect by stimulating the release of melanophore stimulating hormone, responsible for the darkening, and a neurohypophysial hormone, responsible for the induction of the spawning response.

It has been observed that pituitary secretions pass directly to the brain and induce the final stages of maturation and spawning activity. *Gerbilskii*cited in *Pickford and Atz.,* named this secretion the "spawning hormone." Support for the existence of such a hormone was derived from the findings by *Gerbilskii* and other that intracranial injections of pituitary substances were more effective in including final maturation than intraperitoneal or intramuscular injections. However, *Pickford* and *Atz* (1957) point out that other Russian workers have not confirmed these results. *Vivien* (1941) also noted that intracranial implantation of pituitary gland into female *Gobius paganellus* were far more effective in inducing spawning than similar amounts injected into the body cavity. It is impossible at this stage to draw any firm conclusions regarding the mechanism by which spawning is regulated in fishes.

C. Prolactin and Parental Behaviour

Several studies have implicated a prolactinlike hormone in the regulation of parental behaviour in fishes. *Noble et al.* found that injections of prolactin, as well as several other hormone preparations, induced parental behaviour in both males and females of *Hemichromis bimaculatus. Fiedler* (1962) reports that males of the wrasse, *Crenilabrus ocellatus*, treated with prolactin would perform parental fanning even though no nest was present. Similarly, prolactin treatment of *Symphysodon aequifasciata* resulted in a number of changes associated with the parental phase of the reproductive cycle. At the lower dose levels tested there was a marked increase in the number of mucus-secreting cells at the body surface. The secretions of these cells normally provide supplementary nutrient for the young. At the lower dose levels tested there was a marked increase in the number of mucus-secreting cells at the body surface. The secretions of these cells normally provide supplementary nutrient for the young. At high doses the fishes began to perform parental-type fanning directed to a particular place in the aquarium. Agonistic behaviour in response to a test fish was also affected: At intermediate doses there was a decrease fighting behaviour, at higher doses fighting increased. *Blum* and *Fiedler* (1965) examined the effects of pituitary hormone treatments on several species of cichlid including *Pterophyllum scalare, Aequidens Latifrons, Cichlasoma severum,* and *Astronotus ocellatus.* In all cases prolactin treatment resulted in an increase in mucus cell production. This effect was less marked than in *Symphysodon* which is the only species in which the mucus serves as a nutrient for the young.

Some behavioural effects of prolactin treatment were noted. In the case of *Pterophyllum* a low dose a hormone induced a parental type of fanning directed toward a fixed point in space; at higher doses fanning decreased. Agonistic activity in response to a stimulus animal was depressed. Agonistic activity in response to a stimulus animal was depressed by prolactin treatment in *Pterophyllum* and *Aequidens* but unaffected in the other species Prolactin tended to make fish very calm and to depress the feeding response. *Blum* and *Fiedler* regard this latter effect as a component of parental behaviour which prevents eggs or young being eaten by the parents. They also note that hormone-induced parental responses occurred in fish which had had no previous experience of breeding. *Blum* (1966) tested progesterone and several adenohypophysial hormones on *Pterophyllum* and *Symphysodon.* Only prolactin treatment resulted in the appearance of parental behaviour, although both somatotropin and thyrotropin elicited an increase in mucus cell production. Luteinizing and follicle-stimulating hormones resulted in an increase in aggressive behaviour. *Blum* (1966) concludes that prolactin is the only hormone involved in the control of parental behaviour in cichlids.

Prolactin treatment resulted in an increase in mucus cell production in the paradise fish, *Macropodus opercularis.* This effect was more pronounced in males than in females which do not normally build bubble nests or show parental care. However, prolactin treatment alone did not produce an increase in nestbuilding even though the fish produced more mucus. Chorionic gonadotropin alone, or in combination with prolactin, increased nestbuilding behaviour. Similarly, methyltestosterone combined with prolactin treatment resulted in extensive nestbuilding in both males and females. Machemer and *Fiedler* (1965) conclude that for full nestbuilding activity, but this effect is only expressed fully in the presence of a prolactin-induced increase in mucus production.

Smith and Hoar (1967) studied the effects of prolactin treatment on parental and displacement fanning behaviour in the stickleback. They conclude that there is no evidence of a pituitary regulated control of fanning behaviour in the stickleback. In fact, injections of Prolactin appeared to depress fanning, although in no case was the difference between the treatment groups statistically significant.

It is difficult to draw any definite conclusions regarding the role of prolactin in reproductive behaviour in fishes. It is important to separate two results in an increase in mucus production. Second, evidence suggests that prolactin, or rather a prolactinlike hormone, is involved in the regulation of parental behaviour.

The situation ins complicated by the fact that until now only mammalian prolactin has been used in treatment of fishes. Thus differences among various species of fish might result in part from differences in specificity in responsiveness to tetrapod hormone.

EXTERNAL FACTORS AND THE ENDOCRINE SYSTEM

The role of photoperiod, temperature, and other physical factors in the regulation of seasonal changes in reproductive behaviour has been studied. Little is known of the internal mechanisms by which the final the final synchronization is achieved. It is possible that there is no endocrine involvement other than that the sex hormones induce a state of responsiveness to certain classes of behavioural stimuli. On the other hand, it seems more likely that, apart from immediate behavioural responses, the performance of reproductive behaviour may induce endocrine changes in the performing and responding animals, providing the appropriate conditions for the later phases of reproductive activity. In other words, it seems reasonable to look for a situation in fish comparable to that described most clearly in the canary and the ring dove. Not only do hormones affect behaviour, but behaviour in tun affects the endocrine state. The interaction of these relationships results in the endocrine state. The interaction of these relationships results in the integration of behaviour with other physiological conditions and the smooth progression through successive stages of the reproductive cycle.

Relatively little attention has been directed to the details of the breeding requirements of fishes. Aronson (1965) points out that it is commo aquarium practice to lower the temperature of the water to induce cyprinids and characins to breed in captivity. *Chaudhuri* (1960) remarks that in several species of Indian cyprinid spawning only occurs on overcast days and appears to be inhibited by direct sunlight. *Yamazaki* (1965) points out that goldfish will only spawn in the presence of green plants. Bitterling require the presence of a freshwater mussel before they will show any reproductive behaviour. Courtship plays an important role in the final synchronization of the sexes. It seems likely, although there is virtually no experimental evidence, that courtship affects the endocrine state of one or both partners. *Johns et al.* (1969) found that 2-3 days of exposure to a nestbuilding male is necessary before a female gourami will spawn. This suggests that the exposure to the male affects the final maturation and preparation of the of the endocrine state before mating. *Egami* and *Nambu* (1961) maintain that in *Oryzias* male courtship stimulates neurohypophysial activity which in turn is responsible for the induction of oviposition behaviour in the female.

Exposure to other individuals of the same sex may also be important. *Aronson* (1951) notes that females, went through seven or eight spawnings a year. *Aronson* (1965) refers to work involving salmon, herring, whitefish, and minnows which suggests that social facilitation may occur in the synchronization of breeding.

It has been observed. That in the three-spined stickleback, males in winter condition came into breeding more readily if paired, one one each side of a glass partition, than did solitary males. The presence of vegetation also had a stimulating effect upon nestbuilding. On the other hand, the presence of males in breeding condition may have an inhibitory effect upon those without nests. The mechanism of these inhibitory effect upon those without nests. The mechanism of these inhibitory and stimulatory effects is unknown but could involve the endocrine system.

In conclusion, although much is known regarding the proximate environmental factors involved in the long-term control of breeding, little is known regarding the short-term synchronization of sexual partners and the regulation of the successive subcycles which constitute reproductive behaviour in many species of fish. The little evidence available suggests that we may find that endocrine mechanisms are involved and that, as in other vertebrates, we will find that not only do hormones affect behaviour but in turn behaviour affects the endocrine system.